BK 668.4 S521M
MODERN PLASTICS TECHNOLOGY
1975   14.95   FV

3000 348317 30018
St. Louis Community College

668.4 S521m   FV
SEYMOUR
MODERN PLASTICS TECHNOLOGY
            14.95

WITHDRAWN

JUNIOR COLLEGE DISTRICT
of St. Louis - St. Louis County
LIBRARY
5801 Wilson Ave.
St. Louis, Missouri 63110

Modern Plastics Technology

Modern Plastics Technology

RAYMOND B. SEYMOUR

Professor and Coordinator of Polymer Research
University of Houston
Houston, Texas

Adjunct Professor of Polymer Science
University of Southern Mississippi
Hattiesburg, Mississippi

Reston Publishing Company, Inc., *A Prentice-Hall Company*
Reston, Virginia 22090

Library of Congress Cataloging in Publication Data
Seymour, Raymond Benedict, 1912-
   Modern plastics technology.

   Includes index.
   1. Plastics. I. Title.
TP1120.S47   668.4   75-6845
ISBN 0-87909-500-8

MODERN PLASTICS TECHNOLOGY
Raymond B. Seymour

© 1975 by Reston Publishing Company, Inc.
A Prentice-Hall Company
Reston, Virginia 22090

All rights reserved. No part of this book may be
reproduced in any way, or by any means,
without permission in writing from the publisher.

10 9 8 7 6 5 4 3 2 1

Printed in the United States of America

Contents

Foreword   *xi*

Preface   *xiii*

1   **Introduction to Plastics**   **1**

    1.1   Macromolecules or Polymers, 1
    1.2   Plastics, 4
    1.3   Molecular Structure, 5
    1.4   Conformations of Giant Molecules, 5
    1.5   Tacticity, 7
    1.6   Crystallinity, 9
    1.7   Chain Branching, 10
    1.8   Copolymers, 11
    1.9   Transitions in Polymers, 12
    1.10   Cross-linking, 13

CONTENTS

- 1.11 Fillers and Plasticizers, 14
- 1.12 Intermolecular Attractions, 15
- 1.13 Molecular Weights, 18
- 1.14 Backbone Structures, 19
- 1.15 Bulky Groups, 21
- 1.16 The Effects of Solvents, 21
- 1.17 The Effect of Corrosives, 22
- 1.18 Aging of Plastics, 23
- 1.19 The Effect of Heat, 26
- Behavioral Objectives, 27
- Glossary, 28
- Questions, 30
- Answers, 31

2 Compounding Ingredients, Initiators, Chain Transfer Agents, and Fillers   33

- 2.1 The Usefulness of Additives, 33
- 2.2 Initiators, 34
- 2.3 Chain Transfer Agents, Modifiers, Peptizers, 41
- 2.4 Reactive Additives (Curing Agents), 45
- 2.5 Colorants and Pigments, 48
- 2.6 Fillers and Reinforcements, 49
- Behavioral Objectives, 58
- Glossary, 60
- Questions, 61
- Answers, 62

3 Stabilizers, Plasticizers, and Other Additives   64

- 3.1 Antioxidants, 64
- 3.2 Heat and Ultraviolet Stabilizers, 69
- 3.3 Plasticizers, 71
- 3.4 Flame Retardants, 74
- 3.5 Miscellaneous Additives, 76
- Behavioral Objectives, 77
- Glossary, 78
- Questions, 79
- Answers, 79

4 Introduction to Plastics Technology   80

- 4.1 Mixing of Resins and Additives, 80
- 4.2 Casting: Hot Melts, Plastisols, 82
- 4.3 Cellular Plastics, 84

CONTENTS   vii

    4.4   Reinforced Plastics and Laminates, 88
    4.5   Introduction to Rheology, 91
            Behavioral Objectives, 97
            Glossary, 98
            Questions, 100
            Answers, 100

5   *Molding, Extrusion, and Calendering*   *102*

    5.1   Compression Molding, 102
    5.2   Extrusion, 105
    5.3   Blow Molding, 109
    5.4   Injection Molding, 109
    5.5   Powder Molding, 111
    5.6   Calendering, 111
    5.7   Thermoforming, 112
            Behavioral Objectives, 112
            Glossary, 113
            Questions, 114
            Answers, 114

6   *Testing and Characterization of Plastics*   *115*

    6.1   Test Specimens, 116
    6.2   Conditioning of Test Specimen, 116
    6.3   Environmental Tests, 116
    6.4   Dimensional Tests, 118
    6.5   Thermal Tests, 119
    6.6   Mechanical Tests, 120
    6.7   Hardness, 123
    6.8   Electrical Tests, 124
    6.9   Optical Tests, 125
    6.10   Nondestructive Testing, 126
    6.11   Characterization, 127
    6.12   Spectroscopy, 133
    6.13   Thermal Analysis, 134
            Behavioral Objectives, 136
            Glossary, 137
            Questions, 138
            Answers, 138

7   *Phenolic and Other Thermosetting Plastics*   *140*

    7.1   Phenolic Plastics, 140
    7.2   Urea Plastics, 147
    7.3   Melamine Plastics, 149

7.4 Alkyd Plastics, 150
7.5 Unsaturated Polyester Plastics, 153
7.6 Epoxy Resins, 154
7.7 Allyl Plastics, 158
7.8 Silicones, 158
Behavioral Objectives, 160
Glossary, 161
Questions, 162
Answers, 163

8 Polyolefins   164

8.1 Low Density Polyethylenes, 164
8.2 High Density Polyethylene, 166
8.3 Polypropylene, 168
8.4 Olefin Copolymers, 169
8.5 Other Polyolefins, 172
8.6 Miscellaneous Polyolefin Plastics, 174
Behavioral Objectives, 174
Glossary, 175
Questions, 176
Answers, 177

9 Polystyrene and Related Polymers   178

9.1 Polystyrene, 178
9.2 Polystyrene Blends, 180
9.3 Copolymers of Styrene, 181
9.4 ABS Plastics, 184
9.5 Polymers of Styrene Derivatives, 185
9.6 Aromatic Hydrocarbon Polymers, 187
Behavioral Objectives, 190
Glossary, 190
Questions, 191
Answers, 191

10 Poly(Vinyl Chloride) and Related Polymers   192

10.1 Rigid Poly(Vinyl Chloride) (PVC), 192
10.2 Plasticized Poly(Vinyl Chloride), 195
10.3 Copolymers of Vinyl Chloride, 197
10.4 Polymers of Vinylidene Chloride, 198
10.5 Fluoroplastics, 199
10.6 Chlorinated Polyether, 202
Behavioral Objectives, 203
Glossary, 204
Questions, 205
Answers, 205

CONTENTS

11 Saturated Polyesters — 206

11.1 Polymethacrylates, 206
11.2 Polyacrylates, 209
11.3 Polycarbonates, 211
11.4 Poly(Vinyl Carboxylates), 212
11.5 Poly(P-Oxybenzoate), 214
11.6 Polyphthalates, 214
        Behavioral Objectives, 216
        Glossary, 216
        Questions, 217
        Answers, 217

12 Polyamides, Polyimides, and Polyurethanes — 218

12.1 Polyacrylamide, 219
12.2 Nylon-66, 219
12.3 Nylon-6, 222
12.4 Other Polyamides, 223
12.5 Polyamides, 224
12.6 Polyurethanes, 225
12.7 Polyureas, 228
        Behavioral Objectives, 228
        Glossary, 229
        Questions, 230
        Answers, 230

13 Polynitriles, Polyacetals, and Polyalcohols — 231

13.1 Polyacrylonitrile, 231
13.2 Copolymers of Acrylonitrile, 234
13.3 Poly(Phosphonitrilic Chloride), 234
13.4 Polyacetals, 235
13.5 Poly(Vinyl Alcohol), 237
13.6 Poly(Vinyl Acetals), 237
13.7 Polymeric Ethers, 238
13.8 Polyvinylpyrrolidone, 239
        Behavioral Objectives, 240
        Glossary, 241
        Questions, 242
        Answers, 242

14 Starch and Cellulosics — 243

14.1 Starch, 243
14.2 Cellulose, 244
14.3 Inorganic Esters of Cellulose, 246

14.4 Organic Esters of Cellulose, 247
14.5 Cellulose Ethers, 248
Behavioral Objectives, 249
Glossary, 250
Questions, 251
Answers, 251

15 Ablative and Heat Resistant Plastics 252

15.1 Heat Resistant Plastics, 252
15.2 Ablative Plastics, 253
15.3 Polyphenylenes, 254
15.4 Polybenzimidazole (PBI), 254
15.5 Polyquinoxaline, 254
15.6 Polypyrazine, 255
15.7 Polybenzoxazole, 255
15.8 Polybenzothiazole, 255
15.9 Polybenzoxadiazole and Polybenzotriazole, 256
15.10 Polyhydantoin, 256
15.11 Polyketo Polytriketoimidazolidine, 256
15.12 Pyrrone, 257
15.13 Polyanthroline (BBL), 257
15.14 Spiropolymers, 257
Behavioral Objectives, 257
Glossary, 258
Questions, 258
Answers, 259

Index 260

Foreword

Persons entering the plastics field today find themselves in a dynamic growth industry. Few industries in the nation can compare with the remarkable growth of plastics during the last 20 years. This growth has created many challenging jobs requiring the individual to pursue activities which have a significant influence on management decisions. This experience provides a firm background for opportunities to move into management or research and development positions.

The demand for trained personnel to fill these positions is well-documented. The plastics industry is presently active in seeking qualified men and women for job openings in such areas as production, research and development, quality control and sales and service.

One reason for this demand for personnel is that formal education at all levels has not kept abreast of the rate of growth of this industry. This volume is a successful effort to meet the need for other instruments of learning so that technical personnel can readily obtain access to the technological knowledge they need.

Recognizing this need, Dr. Seymour has written a textbook on Modern Plastics Technology in language that is easily understood and which covers the known technology to date—a technology which has undergone extensive changes in the past two decades.

During this period, Dr. Seymour was associated in a very active way with industry and its demands, enabling him to become thoroughly familiar with all significant aspects of industry problems in plastics technology. In this timely book, he presents the fundamental aspects of the subject matter and weaves them into a coherent network that is clear in its meaning and relevant to the reader.

I am happy to recommend this book to both industrial and academic personnel seeking an incisive review of modern plastics technology as well as specific information of particular practical interest.

ALBERT SPAAK
Executive Director
PLASTICS INSTITUTE OF AMERICA

Preface

It has been estimated that over 20 percent of all engineers and technologists are working in some phase of plastics technology. More than 40 percent of all scientists are employed in this field. Since plastics is one of the fastest growing industries, the opportunities for plastics engineers and technologists are increasing and will continue to increase. Formal courses of study are not always available, but every modern technologist, engineer, and scientist must have some knowledge of plastics technology.

This book has been written in an attempt to provide appropriate background information for those who have not taken formal courses in plastics, science, or technology. *Modern Plastics Technology*, in manuscript form, has been used as a textbook for formal courses and as a reference by those working in the field of plastics.

A sincere attempt has been made to write a modern book that will not become outdated, providing the reader consults subsequent relevant articles in plastics technology journals such as *Modern Plastics, Plastics Technology,* and *Plastics World,* as well as encyclopedias issued annually by publishers in plastics technology. In spite of its interest to the general reader, we have not attempted to include much information on the history of plastics: our emphasis has been on plastics technology.

Although plastics is but one phase of polymer technology, we have not attempted here to write a book on polymer technology; hence, little information is provided on rubber, fibers, adhesives, and coatings. In spite of the importance of polymer science, only that science that is essential to the understanding of plastics technology has been included.

In keeping with the technology theme, we have not included literature references nor book lists. The reader who needs more information on any specific plastic should consult the latest editions of the *Encyclopedia of Polymer Science and Technology* and *Modern Plastics Encyclopedia.*

It is our sincere hope that we have written a book that will prove to be of value to plastic technologists. The manuscript was typed by Sharon K. Simmons.

RAYMOND B. SEYMOUR

chapter 1

Introduction to Plastics

1.1 MACROMOLECULES OR POLYMERS

The world around us has always contained polymers. However, prior to World War II, most polymeric materials were naturally occurring, whereas, today, they consist of both natural and synthetic polymers. While the modern polymer scientist understands the physics and chemistry of these essential materials, this was not true of scientists at the time of World War I.

Less than a half century ago, Hermann Staudinger was told by his uninformed colleagues that "there can be no such thing as a macromolecule," i.e., giant molecule.* Presumably, these critics were unaware that they, as all other people, were dependent on naturally occurring biopolymers for protein and starch in their bodies and in their food. Nor were they aware of

---

*R. Olby, *J. Chem. Educ. 47*, 168 (1970).

their dependence on polymeric fibers such as cellulose, wool and silk for their clothing and shelter. For centuries man has been dependent on polymeric paints and coatings for decorative and protective purposes and since the beginning of the twentieth century he has been dependent on polymer rubber tires for much of his transportation.

While the annual world production of all synthetic polymers was less than 100 million lbs at the time of Staudinger's accurate description of polymers in the 1920s, production had grown to almost 10 billion lbs by the time this pioneer polymer chemist received the Nobel prize in 1953. In 1972, the world production of synthetic polymers was approximately 90 billion lbs.

As shown in the accompanying Tables 1-1, 1-2, and 1-3, the total production of plastics, fibers, and elastomers (rubber) in the USA in 1972 was over 17 thousand metric tons, consisting of 11.6 thousand metric tons of synthetic plastics, 3.1 thousand metric tons of fibers and 2.4 thousand tons of elastomers. A total of 875 million gal of paint was also produced in the USA in 1971.

TABLE 1-1 Plastics Sales in the USA in 1972 and 1973 (Thousands of Metric Tons)

Material	1972	1973
Polyethylenes	3,434	3,918
LDPE and Copolymers	2,372	2,664
HDPE	1,062	1,254
Vinyls	2,345	2,462
PVC and Copolymers	1,975	2,171
Other	370	391
Polystyrene and Copolymers	2,111	2,407
Phenolics	651	654
Polypropylene and Copolymers	766	978
Ureas and Melamines	411	464
Alkyds	290	342
Polyesters	416	496
Coumarone-indene resins	137	157
Cellulosics	75	77
Epoxies	83	99
Acrylics	208	233
Polyurethane Foam	493	593
Polyacetals	26.5	31.4
Polycarbonate	25.3	30.9
Nylon	66.5	79.8
Total	11,597	13,182
(Total Production)	11,714	12,273

SOURCE: U.S. Tariff Commission

TABLE 1-2 American and Worldwide Consumption of Fibers in 1972 (Thousands of Metric Tons)

	USA	World
Synthetic Fibers		
Polyester	1,058	2,513
Nylon	898	2,429
Polyacrylic	285	1,272
Polyolefin	188	521
Man-made Cellulosics		
Rayon	439 ⎱	3,540
Cellulose Acetate	195 ⎰	
Natural Fibers		
Cotton	1,691	12,855
Wool	105	1,513
Total Man-made	3,133	10,275
Total Natural	1,796	14,368
Total Man-made and Natural	4,929	24,643

SOURCE: Textile Organon

TABLE 1-3 Elastomer Consumption in the USA in 1972 and 1973 (Thousands of Long Tons)

Synthetic Rubber	1972	1973 (estimated)
Styrene-butadiene	1,440	1,475
Butadiene	305	312
Neoprene	166	180
Butyl	120	144
Isoprene	105	115
Acrylonitrile-butadiene	64	60
Ethylene-Propylene	61	90
Silicone	9.8	10
Urethane	9.0	9.5
Polysulfide	8.7	9.0
Acrylic	6.5	7.0
Total Synthetic Rubber	2,383.2	2,411.5
Total Natural Rubber	630.0	695.0
Total Synthetic and Natural Rubber	3,013.2	3,106.5

SOURCE: U.S. Tariff Commission

It is of interest to note that: 10 of the 15 recent major breakthroughs associated with the petrochemical industry were related to the producton of polymers, the volume of synthetic polymers exceeds that of all other chemicals, and more scientists and technologists are employed in polymer-related jobs than in any other field.

The general term *polymer* includes all natural and synthetic plastics, fibers, elastomers, paints, and adhesives. Prior to World War II, natural rubber and natural fibers were used almost exclusively. However, as shown in Tables 1-2 and 1-3, the production of synthetic rubber and fibers in the USA exceeds that of natural products. Since there are few natural occurring plastics, plastics production is almost exclusively devoted to synthetic plastics. In spite of the importance of other types of macromolecules, the subsequent discussions in this book will be limited to plastics.

1.2 PLASTICS

A *plastic* is any kind of matter that flows as the result of the application of heat and/or pressure. The American Society for Testing Materials (ASTM) has defined plastics as "materials that contain as an essential ingredient organic substances of high molecular weight which are solid in the finished state but are shaped by flow at some stage of their manufacture or during processing into finished articles."

Since most plastics are organic substances, they contain the element carbon (C). The carbon atoms present in these polymer molecules are joined with carbon atoms or other atoms such as oxygen (O) or nitrogen (N) in continuous chains that are giant molecules. If one could magnify these macromolecules so that they were visible to the eye, one would see that they look like long pieces of twisted wire. In contrast an ordinary molecule such as propane $(H(CH_2)_3H)$ which is present in natural gas would look like a bent pin. Simulated structures of ethane and high-density polyethylene $(H(CH_2)_{2000}H)$ are illustrated in Fig. 1-1 below.

Propane   Polyethylene

Fig. 1-1 Magnified simulated structures of ordinary molecules and macromolecules.

1.3 MOLECULAR STRUCTURE

The characteristic angle of 109.5° shown for the propane structure in Fig. 1-1 would be repeated two thousand times in the macromolecular structure but was omitted in order to present a simpler diagram. Actually, a more correct model of a polyethylene molecule would be a kinked and twisted piece of barbed-wire fencing. The kinks at an angle of 109.5° would represent the characteristic bond angle between each carbon atom in the chain and the barbs would represent the two hydrogen (H) atoms attached to each carbon atom at the same tetrahedral angle (109.5°).

These hydrogen atoms, also present on the carbon atoms of ethane, are of considerable interest to the organic chemist. However, the plastics technologist is much more interested in the twists that are characteristic of the higher molecular weight molecules. Polymer chains consisting of repetitive methylene (CH_2) groups are said to be linear because their length is at least a thousand times greater than their thickness.

However, as evident from Fig. 1-1, these macromolecules seldom exist as straight rods. A magnified short section of this carbon-carbon chain would have the zigzag arrangement shown for ethane. The distance between the centers of carbon atoms along the bond angles would be 1.54 angstroms (Å) where 1 angstrom is equal to one hundred-millionth of a centimeter (1×10^{-8} cm).

The distance between the centers of each carbon atom along the direction of the chain would be somewhat less than 1.54Å, or 1.26Å. Thus, the chain shown in Fig. 1-1 would have a length of 2520Å if it were stretched like a barbed wire fence between two poles. However, these chains are seldom extended to their full contour length but are twisted and coiled as a result of the rotation of the carbon-carbon bonds in the chain or backbone of the polyethylene molecule.

1.4 CONFORMATIONS OF GIANT MOLECULES

While most plastics and other polymers are rigid at room temperature, the polymer chains in flexible plastics are actually wiggling in a snakelike fashion. As a result, these chains are present in many different coil-like shapes or conformations. Thus, one seldom considers the fully stretched out or full-contour length ($n\ell$) of a polymer chain but employs a statistical approach that measures the average end-to-end distance (r) for the many possible random coil conformations. This average is a reasonably good estimate of the size of the molecule.

The statistical approach is patterned after a classical calculation solved by Lord Rayleigh over a century ago. This approach, called the *random flight technique*, yields a value that is essentially the linear distance from start to finish of a blindfolded person taking n number of steps of length ℓ. This distance, called the *root mean square distance* ($\sqrt{\bar{r}^2}$), is equal to $\ell\sqrt{n}$.

According to this formula, the average end-to-end distance for a linear polyethylene molecule consisting of 2,000 methylene groups (CH_2) would be $1.54\sqrt{2000}$Å or 69Å in contrast to the previously calculated, full-contour length of 2520Å. The calculated average end-to-end distance is less than 3% of the full-contour length.

The value obtained by the Raleigh approach for the end-to-end distance is less than the actual length of a polymer chain because the polymer chain does not have the freedom of motion possessed by the blindfolded walker. The classical formula ($\ell\sqrt{n}$) assumes freedom of motion not possible for a carbon-carbon chain having a fixed-bond angle of 109.5°. If one neglects the effect of the hydrogen atoms on the polyethylene chain, the corrected value of the chain length is about 30% greater than the value of 69Å or about 98Å.

When one corrects for the restricted motion or hindrance caused by the presence of the hydrogen atoms on the chain, one achieves the new value for the end-to-end distance of 122Å. There is also an additional hindrance to free rotation because of what Flory has termed *pentane interference*. This is characteristic of a simple linear hydrocarbon called pentane ($H(CH_2)_5H$), in which rotation is hindered by an overlapping of the hydrogen atoms on the terminal carbon atoms of this simple molecule.

When the Flory correction is applied to the polyethylene molecule, it shows that the molecule's average end-to-end distance increases to about 180Å. This is approximately 2.5 times the calculated value of 69Å.

There is still another correction necessary for the calculation of the true average end-to-end distance of a real polymer chain. It is related to the existence of a so-called *excluded volume*. The blindfolded walker may backtrack without interference. However, only one atom in the carbon chain can occupy any one volume at the same time and all other atoms in the chain must be excluded. Unfortunately, no one has proposed a useful correction factor to correct for this excluded volume.

Nevertheless, the value of 180Å is a good approximation and until one learns how to correct for the excluded volume it will continue to be used to estimate the end-to-end distance for a chain of 2,000 carbon atoms. It is important to note that the ratio for the actual average end-to-end distance to the uncorrected value obtained by the random flight method is a measure of the stiffness of the polymer chain.

1.5 TACTICITY

The flexibility of a plastic such as polyethylene will increase as the temperature is increased from room temperature to the temperature of boiling water because of the increased mobility of the polymer chains. In contrast, this flexibility at any specific temperature will decrease if properly-ordered pendant groups, such as methyl groups (CH_3), are present in place of some of the hydrogen atoms on the polymer chain or backbone. Thus, the physical properties of polypropylene (PP), which has a pendant methyl group on every other carbon atom, will differ from those of polyethylene, and these properties will depend on the relative positions of these pendant groups.

Natta, who shared the Nobel Prize with Ziegler in 1963, described the principal different arrangements of these groups as *tacticity*, a term derived from the Greek word *taktika*, meaning order. Accordingly, when the pendant methyl groups are present on every other carbon atom and are all on the same side of the chain, the structure is said to be *isotactic*.

When every other pendant group is on the opposite side of the polymer chain in an alternating configuration, the structure is said to be *syndiotactic*. The term *atactic* or *heterotactic* is used to designate randomness or lack of order in the arrangement of the pendant groups, in contrast to the isotactic and syndiotactic structures which are said to be *stereoregular*.

Because of their inherent ordered structures, the stereoregular polymer chains fit together well, resulting in a tendency for crystallization. Thus, isotactic polypropylene is more rigid and has a much higher melting point (T_m) than polyethylene. Skeletal segments of these two polymer chains are shown in Fig. 1-2.

$$-C-C-C-C-C-C-$$

High density (linear)
polyethylene T_m 135°C

$$\begin{array}{cccccc} & & C & & C & & C \\ & & | & & | & & | \\ -C- & C- & C- & C- & C- & C- \end{array}$$

Isotactic polypropylene
T_m 165°C

Fig. 1-2 Skeletal formulas of typical segments of polymer chains.

Skeletal segments of polymer chains for syndiotactic and atactic polypropylene are shown in Fig. 1-3. It is important to observe that in all structures shown for polypropylene, the pendant methyl group is on every other carbon. This structure suggests that each molecule of propylene $\left(\begin{array}{c} CH_3 \\ | \\ H_2C{=}CH \end{array} \right)$ adds to the growing chain in the same manner during polymer formation, i.e., polymerization. The arrangement of these segments or mers forms a

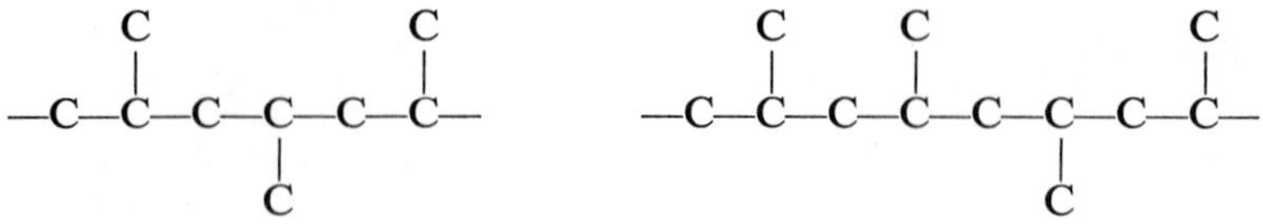

```
      C         C                    C     C            C
      |         |                    |     |            |
   —C—C—C—C—C—C—                  —C—C—C—C—C—C—C—C—
            |                             |
            C                             C
```

    Syndiotactic polypropylene   Atactic polypropylene

Fig. 1-3   Skeletal formulas for polypropylenes.

head-to-tail configuration. This arrangement, and not a head-to-head configuration, is typical of most polymer chains. These structures are shown by skeletal formulas in Fig. 1-4.

    Because of the bulkiness of the pendant methyl group, the volume of the polypropylene chain is greater than that of a linear polyethylene chain of comparable length. Since density and volume are related inversely, the density of polypropylene (0.90 g/cm^3) is less than that of linear polyethylene (0.95 g/cm^3).

    As stated previously, the melting point of polypropylene is greater than that of polyethylene. However, when a larger linear group such as ethyl (C_2H_5) or propyl (C_3H_7) is present instead of the methyl group, as in polybutene and polypentene, the melting points are decreased to 125 and 75°C respectively. The densities of these polymers are also less than polypropylene because of the volume occupied by the bulkier groups which permit the backbones of these chains to wiggle more than more closely-packed chains.

    Providing there is no sacrifice in other properties, such as the melting point, low density is an advantageous property, since plastics and other materials of construction are used on the basis of volume rather than weight. Fortunately, when a branched chain, such as an isobutyl group ($-CH_2-CH-(CH_3)_2$), or a long linear chain, such as a decyl group ($-(CH_2)_{10}H$), is present as the pendant group the melting point of the polymer is increased.

    Poly(4-methylpentene-1) (TPX), which has a density of 0.83 g/cm^3 and a melting point of 300°C, is commercially available. Presumably the branched chain present in this isotactic polymer reduces the wiggling of the polymer backbone. The increased melting points of the polymers with very large linear bulky groups are assumed to be related to the length of these groups. These regularly spaced sidechains are mutually attracted in much the same way that linear polyethylene chains are attracted to each other.

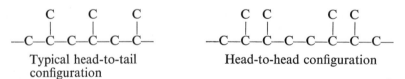

    Typical head-to-tail   Head-to-head configuration
    configuration

Fig. 1-4   Skeletal formulas for head-to-tail and head-to-head configurations of polypropylene.

1.6 CRYSTALLINITY

The increased rigidity of stereoregular polymers noted in the previous section is related to their geometry which permits a more orderly arrangement of the molecules. The packing of molecules and their crystallization tendency is comparable to the crystallization of low molecular weight organic molecules, such as cane sugar (sucrose) or benzoic acid. While the crystals of the low molecular weight products may be observed visually, instrumental techniques are required for determining the extent of crystallinity in polymers.

The presence of spherical aggregates of crystals or spherulites causes some polymers to be translucent or opaque at ordinary temperatures. The degree of crystallinity may be estimated by x-ray diffraction, infrared spectroscopy or by ÑMR techniques. Classical studies of the form or morphology of polymers led to the suggestion that both noncrystalline or amorphous and crystalline areas were present in polymers. These were depicted by a fringed-micelle model. Since single crystals of polyethylene have been prepared, the fringed-micelle model is no longer essential for an explanation of the morphology of crystalline polymers.

An examination of the single crystal platelets of polyethylene has provided a new concept of polymer morphology. Since the thickness of these sheet-like crystals or *lamella* is only about 100Å, it is obvious that the thickness could not provide enough room for a polymer chain in that direction. However, electron diffraction data have shown that the polymer chains are not laid out in the direction of the longer axes of these plates. Instead, the lamellae consist of chains folded back and forth throughout the entire thickness of the platelets, as shown in Fig. 1.5.

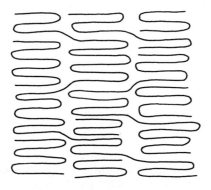

Fig. 1-5   Folded polymer chains in polymer crystals.

It is now believed that a solid polymer such as polyethylene consists of spherulites that are made up of lamellae. The latter consists of folded chains and may be bonded together at many points.

If a film of polyethylene is heated, it becomes more transparent as the crystals melt. However, if the warmed transparent film is cooled slowly, large spherulites will form and the film will be weakened at the boundaries of these spherulites. These defects may be reduced to some extent if a small amount of readily crystallizable organic compound such as benzoic acid is present in the film and the warmed film is chilled rapidly.

Presumably the small crystals of benzoic acid that are formed in the film serve as centers or nuclei for the formation of many small crystals or crystallites. Rapid cooling also minimizes the formation of large spherulites which contribute more to opacity than small crystals.

1.7 CHAIN BRANCHING

As shown in Fig. 1-6, the structure of low density polyethylene (LDPE) is highly branched and much less ordered than the high density polyethylene (HDPE) molecule shown in the same figure. These branches in LDPE resemble the previously discussed pendant groups and, like the latter, cause an increase in volume relative to the unbranched polymer chain. Since the volume is related inversely to the density, the density of a branched polymer such as LDPE is less than that of a linear polymer such as HDPE. The values in this case are approximately 0.91 g/cm^3 and 0.95 g/cm^3. This difference is recognized in the standard LDPE and HDPE nomenclature.

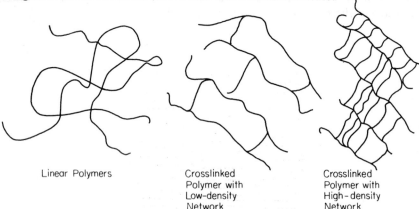

Linear Polymers

Crosslinked Polymer with Low-density Network

Crosslinked Polymer with High-density Network

Fig. 1-6 Sketches illustrating the general structure of linear (HDPE) and branched (LDPE) polyethylene molecules.

Because of the symmetry of its structure, HDPE is a crystalline polymer. In contrast, because of the lack of symmetry resulting from randomly distributed branches of varying length, LDPE is less crystalline. However, the latter is not amorphous at room temperature, as evidenced by the translucency of thick films of LDPE.

The average end-to-end distance may be readily calculated for HDPE, However, because of its complex structure, it is difficult to ascertain which are the true chain ends in a highly branched polymer such as LDPE. Accordingly. one uses the term *radius of gyration* (s), which is equal to the square root of the sum of the square of the distance of various points from the center of gravity of the molecule. The length of the radius of gyration is less than the end-to-end distance (r) of a polymer molecule, and, for a linear molecule r, is equal to about $\sqrt{6S}$ or $2.45S$.

1.8 COPOLYMERS

A high degree of crystallinity such as that observed for polypropylene is an essential property for strong fibers. Crystallinity contributes to the strength and opacity of plastics but is not a desirable property in elastomers (rubbers). As noted previously, the tendency toward crystallization in polymers is decreased by the presence of irregularly spaced bulky pendant groups or branches. Crystallinity may also be decreased by incorporating other building units or mers in the polymer backbone.

Both polyethylene and polypropylene are produced at an annual rate of over a million tons by the polymerization of ethylene and propylene monomers. When a mixture of monomers such as ethylene and propylene is subjected to the same conditions used to produce individual polymers or homopolymers, copolymers are often obtained. Both of the original mers, i.e., ethylene and propylene, are present in the backbone of these copolymers and these macromolecules are usually amorphous rather than crystalline at ordinary temperatures.

The mer units (A and B) are usually randomly distributed in the copolymer and different products may be produced from these same mers by varying the composition of the reactant mixture. However, if one of the mers has little tendency to polymerize by itself, the mers may be arranged in alternating positions in the backbone as shown in Fig. 1-7.

—AABABBAAABA—	—ABABABABAB—
Random copolymer	Alternating copolymer

Fig. 1-7 Segments of random and alternating copolymers of mers A and B.

By use of appropriate polymerization techniques, it is possible to produce copolymers with long sequences of each mer or copolymers in which one of the mers is present as branches on the homopolymer chain. These copolymers, shown in Fig. 1-8, are called block and graft copolymers and have different physical properties than random copolymers.

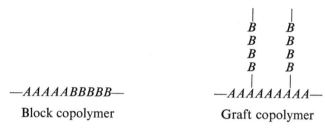

Fig. 1-8 Segments of block and craft copolymers of mers A and B.

Different properties may also be obtained by the physical blending of homopolymers. The most interesting properties are often obtained when the two components of the mixture are incompatible so that one polymer is dispersed in a continuous matrix of the other polymer in the solid state.

1.9 TRANSITIONS IN POLYMERS

Amorphous polymers or copolymers are usually hard, stiff, transparent, and somewhat brittle at low temperatures. They are said to be in the glassy state. These amorphous polymers become more flexible and less brittle as a result of the onset of chain mobility when heated to a characteristic temperature called the glass transition temperature (T_g). At temperatures above T_g, the polymer chain requires more volume for free rotation about its axis. Hence, an abrupt change in volume and density is noted at T_g, which is sometimes called the *second order transition temperature.*

The T_g values are dependent to some extent on the technique used and the time consumed in the determination of these transition temperatures. However, the temperature ranges reported for T_g for plastics are usually above room temperature and are characteristic for each polymer. The T_g values for polystyrene (PS), poly(vinyl chloride) (PVC), and poly(methyl methacrylate) (PMMA) are 100, 82, and 105°C respectively.

When crystalline polymers are heated at relatively high temperatures, melting occurs and these polymers become viscoelastic liquids at their characteristic melting points (T_m). This first order transition (T_m) is also observed when amorphous polymers are heated above their glass transition temperatures. The values for T_m are usually 33 to 100% greater than those observed for T_g. The percentage difference between T_g and T_m is greatest for symmetrical polymers such as polyethylene and least for unsymmetrical polymers such as poly(vinyl chloride).

The T_g value for polypropylene is $-18°C$ or $255°K$, yet because of its high degree of crystallinity, it does not flow excessively below its T_m value of $176°C$. In contrast, the T_g value for polyisobutylene $\left(-\underset{\underset{C}{|}}{\overset{\overset{C}{|}}{C}}-C-\right)_n$ is $-70°C$ and this product will flow slowly at room temperature. While it is not possible to view the segmental motion of amorphous polymer chains at temperatures above T_g, it is possible to observe the effects of this motion. Thus, when a thick sheet of polyisobutylene is allowed to extend over a table edge at room temperature, it will cold flow from the table to the floor over a period of several days.

1.10 CROSS-LINKING

Cold flow is the result of chain mobility which permits one chain to slip by another. It may be overcome by the introduction of crosslinks between the individual polymer chains. The structure present in these network polymers prevents slippage in much the same way that wires used to form a bedspring are not able to move independently.

There is a small degree of crosslinking in naturally occurring protein molecules. This is the result of sulfur-sulfur linkages present in an amino acid building block called *cystine*. However, few natural polymers are crosslinked in their natural state. For example, natural rubber is a sticky product with limited use because of the unrestricted slippage of its polymer chains. Charles Goodyear overcame this deficiency by heating natural rubber with a small amount of sulfur to produce the network polymer which is now used in rubber bands and tire treads.

Butyl rubber is a synthetic copolymer of isobutylene and isoprene $\left(\underset{H_2C=CH-CH=CH_2}{\overset{CH_3}{|}}\right)$. The latter is the minor constituent and its concentration in the reactant mixture is just enough to provide a few double bonds ($-C=C-$) in each polymer chain. Natural rubber is a homopolymer of isoprene and contains many double bonds, all of which are capable of being crosslinked with sulfur.

Charles Goodyear introduced only a few crosslinks by using only a small amount of sulfur. The butyl rubber molecule is tailored so that only a few sulfur crosslinks can be introduced regardless of the amount used for curing or vulcanization.

Both vulcanized butyl rubber and natural rubber that have been cured with a limited amount of sulfur are elastic, because there is ample opportunity for the local movement of chain segments between the occasional crosslinks. However, if natural rubber is heated with a large amount of sulfur, many crosslinks are introduced and the product becomes infusible hard rubber or ebonite. Segmental movement of polymer chains is restricted when there is a high crosslink density, as is the case with hard rubber and many synthetic plastics such as Bakelite.

Since the conversion of a linear polymer to a network polymer is often caused by heating with a crosslinking agent such as sulfur, the process is called *thermosetting* and the products *thermoset plastics*. The linear and branched polymers are softened by heat. Often, they are called *thermoplastics* Appropriate solvents are able to promote chain mobility in thermoplastics and to dissolve these polymers. However, thermosetting plastics may be swollen by solvents but they are insoluble in all solvents. Typical structures for linear and network polymers are shown in Fig. 1-9.

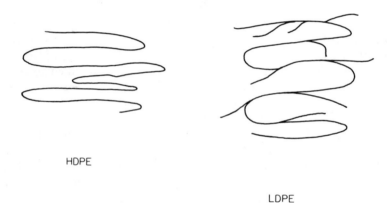

HDPE

LDPE

Fig. 1-9   Linear and network structures.

1.11 FILLERS AND PLASTICIZERS

In contrast to the sticky, unvulcanized, natural rubber, the cured product maintains its elasticity over a wide temperature range and is suitable for numerous applications. However, it is not completely satisfactory for use in tire treads, which is its principal application. The pioneer bicycle and carriage tires were colored black as the result of the addition of a small amount of finely divided carbon.

Early in the 20th century, it was discovered accidentally that natural rubber could tolerate large amounts of carbon black and that this inexpensive additive improved the physical properties of the tire treads. Gum stock tires would not be useful for more than a few thousand miles of service, whereas carbon-black tread stocks have proved satisfactory for over 40 thousand miles of service.

The addition of a filler such as carbon black, silica or asbestos, reduces the mobility of polymer chains and thus decreases stickiness and cold flow. This effect is enhanced when crystalline polymers are used or when a few crosslinks are present in amorphous polymers.

Whereas fillers restrict the movement of polymer chains, compatible solvents increase flexibility of polymers by permitting independent movement of the polymer chains. Nonvolatile compatible liquids are called *plasticizers* since they promote segmental motion and reduce both T_g and T_m in accordance with the amount added.

Poly(vinyl chloride) is a rigid, brittle polymer. A more flexible product may be obtained by the copolymerization of vinyl chloride with other monomers such as vinyl acetate or methyl acrylate. The term *internal plasticization* is sometimes used to describe the effect of such comonomers. When high-boiling liquids such as dioctyl phthalate are added to PVC, flexible plastics are obtained. These high-boiling liquids are sometimes called *external plasticizers*.

1.12 INTERMOLECULAR ATTRACTIONS

A simple hydrocarbon such as ethane ($H(CH_2)_2H$) is a gas at room temperature but it will change to a liquid if cooled below $-90°C$ or if it is compressed at higher temperatures. In contrast, a higher molecular weight hydrocarbon such as hexane $H(CH_2)_6H$ is a liquid at room temperature. However, it will boil, i.e., become gaseous, when heated above 69°C and solidify when cooled below $-95°C$.

Whether a substance is a solid, liquid, or gas depends on the kinetic energy present, which tends to separate the molecules as the temperature is increased. The attractive forces between the molecules that oppose this separation are called secondary valence forces, or Van der Waals' forces. These forces, which are present in all substances, are proportional to their molecular weights. Hence, the intermolecular forces present in a moderately sized molecule such as hexatriacontane $H(CH_2)_{36}H$, present in paraffin wax, are many times greater than those present in hexane. However, because of its extremely high molecular weight, the secondary valence forces in polyethylene are many times greater than those in the constituents of paraffin wax.

The simplest type of dispersion forces, called *London forces*, are present in hydrocarbon molecules such as ethane and polyethylene. The molecules are temporary dipoles as a result of the distortion of the electron clouds around the carbon and hydrogen atoms present. A simple hydrocarbon molecule such as methane is neutral or nonpolar since, on a statistical basis, the positive and negative charges or temporary dipoles are balanced. Nevertheless, since the charges on the electron clouds in methane may be distorted momentarily, these symmetrical molecules are polarized for an extremely small fraction of a second. Thus, the transient dipoles present momentarily act as micromagnets, as shown in Fig. 1-10.

Fig. 1-10 Attraction of temporary dipoles resulting from momentary distortion of electron clouds in nonpolar molecules.

In spite of their transient existence, there are enough temporary dipoles to produce many transient bonds which hold the liquid ethane molecules together at $-90°C$. The temporary dipoles also cause two or more molecules in paraffin wax to be attracted to each other at room temperature and even at slightly higher temperatures. However, because of the short length of the chains in these molecules, paraffin wax is a brittle material.

The secondary valence forces present in polyethylene are many times greater than those in paraffin wax. It is also of interest to note that because of the entanglement of polymer chains, these giant molecules are not brittle. The strength of a hydrocarbon polymer such as polyethylene is based on both the London-type Van der Waals' forces and the entanglement of polymer chains.

The fact that the boiling point of methyl chloride (HCH_2Cl) is over 150°C higher than that of methane (HCH_2H) cannot be accounted for by the increase in temporary dipoles of these molecules only. The major difference in the intermolecular attraction between these molecules is the dipole-dipole interaction of these permanent dipoles.

As shown in Fig. 1-11, these stronger secondary valence forces are also present in many rigid polymers such as poly(vinyl chloride) (PVC) and polyacrylonitrile (PAN). The charges are labelled as δ^+ and δ^-. The high glass transition temperature values of poly(vinyl chloride) and polyacrylonitrile, 87°C and 130°C respectively, are the result of both the London-type and the dipole-dipole intermolecular forces in these molecules.

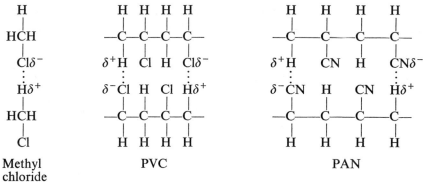

Fig. 1-11 Typical dipole-dipole interaction between molecules of methyl chloride and segments of chains of poly(vinyl chloride) and polyacrylonitrile.

The boiling point of methanol (CH_3OH), which is about 90°C higher than methyl chloride, cannot be accounted for entirely by the previously mentioned secondary valence forces. Actually, there is an additional intermolecular force present in molecules containing hydroxyl (OH) or amino (NH_2) groups. These very strong intermolecular forces, called *hydrogen bonds*, are shown for nylon-66, in Fig. 1-12.

The higher melting point of nylon-66, which is 265°C, is the result of London forces, dipole-dipole interactions, and hydrogen bonding in these molecules. While these forces are much smaller than the primary valence bonds between atoms in molecules, they are additive and hence are strong enough to reduce molecular motion in much the same manner as the previously discussed crosslinks.

The magnitude of these intermolecular forces increases from the weak London forces to the stronger dipole-dipole interaction and to the very strong hydrogen bonds. However, the primary valence bonds are ten times stronger than the hydrogen bonds. It is of interest to note that the melting point (T_m) of

$$
\begin{array}{c}
-\text{C}-(\text{CH}_2)_4-\text{C}-\text{N}-(\text{CH}_2)_6-\text{N}-\text{C}-(\text{CH}_2)_4- \\
\parallel \qquad\qquad\quad \parallel \ \ \mid \qquad\qquad\ \ \mid \ \parallel \\
\text{O}\delta^- \qquad\qquad \text{O} \ \ \text{H}\delta^+ \qquad\quad \text{H} \ \ \text{O}\delta^- \\
\vdots \qquad\qquad\qquad\qquad \vdots \qquad\qquad\qquad \vdots \\
\text{H}\delta^+ \qquad\qquad \text{H} \ \ \text{O}\delta^- \qquad\quad \text{O} \ \ \text{H}\delta^+ \\
\mid \qquad\qquad\qquad \mid \ \parallel \qquad\qquad \parallel \ \mid \\
-\text{N}-(\text{CH}_2)_6-\text{N}-\text{C}-(\text{CH}_2)_4-\text{C}-\text{N}-(\text{CH}_2)_6-
\end{array}
$$

Fig. 1-12 Typical hydrogen bonding between hydrogen and oxygen or nitrogen atoms in nylon-66.

nylon-66 is decreased by 120°C when hydrogen bonding is prevented by replacing the hydrogen atoms on the nitrogen atoms by methyl groups (CH_3).

In special cases such as ionomers, polymer molecules are attracted to each other at ordinary temperatures by strong ionic or electrostatic bonds. The latter forces are reduced at elevated temperatures so that these ionomers may be readily molded and extruded. The ionic bonds present in a segment of an ionomer are shown in Fig. 1-13.

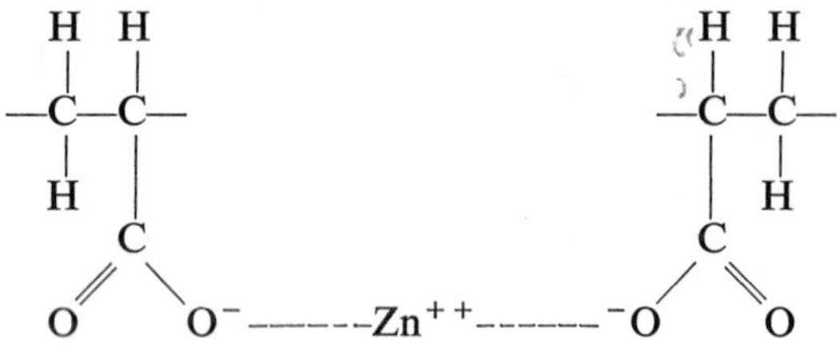

Fig. 1-13 A typical segment showing ionic bonding in an ionomer.

1.13 MOLECULAR WEIGHTS

As noted previously, plastics and other polymer molecules consist of long chains contributing to the characteristic properties of plastics by entanglement and by intermolecular attractions. Actually, a minimum chain length is essential for a molecule to exhibit characteristic polymeric properties. This threshold chain length is shorter for true network polymers than for polymers with strong intermolecular forces. Considerably higher molecular weight values are required before linear molecules exhibit polymeric properties.

Ordinary small molecules such as methanol have one molecular weight only. Other alcohols such as ethanol also have only one molecular weight but the molecular weights of these two homologs differ by 14. They are 32 and 46 respectively. Molecules having one molecular weight only are called *monodisperse* whereas a mixture of homologs such as methanol and ethanol is called *polydisperse*.

Some naturally occurring macromolecules such as proteins are monodisperse, but cellulose, natural rubber, and most synthetic polymers consist of a mixture of different molecular weight molecules or homologs and hence are polydisperse. The molecular weight of an equimolar mixture of methanol and ethanol is 39, which is the average molecular weight of the mixture of these homologs. It is also customary to use the term average molecular weight for polydisperse polymers.

The number of monomer units or mers present in a polymer chain is usually represented by the letter n as shown previously for polyethylene.

The number of mers (n) is also called the *degree of polymerization*, abbreviated as *DP*. Molecular weight is equal to the weight of the mer multiplied by the number of mers (n). The degree of polymerization for a mixture of homologs in a polydisperse polymer is an average DP and is designated by use of an overbar i.e., $\overline{DP}$.

The average degree of polymerization ($\overline{DP}$) is calculated by dividing the sum of the individual *DP* values by the number of molecules present in the mixture. When this procedure is applied to molecular weights, the average is called the *number average molecular weight* and is designated as $\overline{M}_n$. Values for $\overline{M}_n$ are obtained by actually counting the molecules using techniques such as osmometry, discussed in a later chapter.

In some instances, it is preferable in a polydisperse system to emphasize the size of the polymer molecules rather than their number. The weight average molecular weight $\overline{M}_w$ is calculated by adding up the squares of the individual molecular weight values and dividing this sum by the sum of the individual molecular weight values. As discussed in a subsequent chapter, the $\overline{M}_w$ may be obtained by light scattering techniques.

One may separate the individual homologs in a polydisperse system by the gradual addition of a miscible nonsolvent to a solution of the polydisperse system. Similar data may be obtained continuously from gel permeation chromatography (GPC) in which the larger molecules are not as readily entrapped as the smaller molecules and thus pass more easily through a packed column. This process is called *elution* and the liquid passing through the column or *eluate* is richer in high molecular weight fractions while the lower molecular weight fractions are preferentially retained in the column. Subsequent eluates will contain lower molecular weight fractions.

The data for molecular weight distribution may be plotted on a graph, such as that shown in Fig. 1-14. In a typical molecular weight distribution curve, it is customary to show the percentage of total polymer on the vertical axis and the molecular weight values on the horizontal axis. As illustrated in Fig. 1-14, the weight average molecular weight ($\overline{M}_w$) is greater than the number of average molecular weight ($\overline{M}_n$) for a polydisperse polymer. The ratio $\overline{M}_w/\overline{M}_n$ for a monodisperse polymer is 1 and the degree of dispersity for a polydisperse polymer may be determined from this ratio.

1.14 BACKBONE STRUCTURES

Because of the relative ease of rotation of the carbon-carbon bonds in polymers with catenated carbon chains, such as polyethylene, polymer chains are flexible at ordinary temperatures. This flexibility is reduced when stiffening groups, such as phenylene groups $\left(-\!\!\left\langle\bigcirc\right\rangle\!\!-\right)$ are present in the

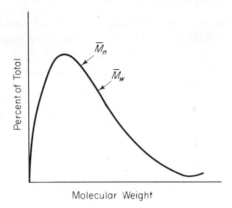

Fig. 1-14 A typical curve for the molecular weight distribution of a polydisperse polymer.

backbone. Thus, while polyethylene $\left(\begin{array}{cccc} H_2 & H_2 & H_2 & H_2 \\ | & | & | & | \\ -C-C-C-C- \end{array}\right)$ has a T_g value of $-125°C$, the T_g value for poly(p-xylylene) $\left(-\overset{H_2}{\underset{|}{C}}-\underset{}{\bigcirc}-\overset{H_2}{\underset{|}{C}}-\right)$ is 70°C.

Because of the ease of rotation of carbon-carbon chain segments, the flexibility of the polymer is increased and the T_g value is decreased when additional methylene groups $\left(\begin{array}{c} H_2 \\ | \\ -C- \end{array}\right)$ are present between phenylene groups.

Comparable effects are also noted when other atoms besides carbon are present in the backbone. Thus, poly(ethylene terephthalate) $\left(-\overset{O}{\underset{\|}{C}}-\bigcirc-\overset{O}{\underset{\|}{C}}-O-\overset{H_2}{\underset{|}{C}}-\overset{H_2}{\underset{|}{C}}-O-\right)$ has much higher T_g and T_m values than poly(ethylene adipate) $\left(-\overset{H_2}{\underset{|}{C}}-\overset{H_2}{\underset{|}{C}}-O-\overset{O}{\underset{\|}{C}}-\overset{H_2}{\underset{|}{C}}-\overset{H_2}{\underset{|}{C}}-\overset{H_2}{\underset{|}{C}}-\overset{H_2}{\underset{|}{C}}-\overset{O}{\underset{\|}{C}}-O-\overset{H_2}{\underset{|}{C}}-\overset{H_2}{\underset{|}{C}}-\right)$.

Likewise, aromatic polyamides $\left(-\overset{O}{\underset{\|}{C}}-\bigcirc-\overset{O}{\underset{\|}{C}}-\overset{H}{\underset{|}{N}}-(CH_2)_n-\overset{H}{\underset{|}{N}}-\right)$ also have higher T_g and T_m values than the corresponding aliphatic polyamides. $\left(-\overset{O}{\underset{\|}{C}}-(CH_2)_n-\overset{O}{\underset{\|}{C}}-\overset{H}{\underset{|}{N}}-(CH_2)_n-\overset{H}{\underset{|}{N}}-\right)$.

1.15 BULKY GROUPS

The T_g and T_m values for isotactic and syndiotactic polymers are increased when regularly spaced bulky groups are present. However, these values are decreased when bulky groups are present in atactic polymers. Thus, while atactic poly(methyl methacrylate) has a T_g value of 105°C, the values for the corresponding polyethyl, poly(n-propyl), and poly(n-butyl methacrylates) are 47°, 33°, and 17°C respectively.

These T_g values continue to decrease as the size of the ester group increases from five to ten carbons. However, because of side-chain crystallization between linear side chains, this trend is reversed when more than ten carbon atoms are present in the ester group. Thus, the T_g values are −70 and −9°C for poly(n-decyl methacrylate) and poly(n-tetradecyl methacrylate) in which the ester groups contain 10 and 14 carbon atoms respectively.

1.16 THE EFFECT OF SOLVENTS

When polymers are to be used as protective coatings or plastic cements, it is customary to select a solvent whose intermolecular attractions are similar to those present in the polymer molecules. The opposite approach is used when solvent resistance is required. These intermolecular forces, called *cohesive energy density* (CED), are actually the energy required to evaporate one mole of solvent. In practice, the square root of the cohesive energy density is used. This square root value is called the *solubility parameter* and is represented by the symbol δ using hildebrand units (H).

The solubility parameter value, which is actually a measure of the polarity of a solvent, is also applied to polymers even though it is impossible to determine their molar energy of vaporization. The δ values for polymers are estimated by observing the degree of swelling of crosslinked polymers in solvents with known δ values.

Solubility parameter values range from 6.2H for polytetrafluoroethylene $\left(\begin{array}{cc} F_2 & F_2 \\ | & | \\ -C-C- \end{array} \right)$ and neopentane $\left(H_3C-\underset{\underset{CH_3}{|}}{\overset{\overset{CH_3}{|}}{C}}-CH_3 \right)$ to 15.4H for polyacrylonitrile $\left(\begin{array}{cc} H_2 & H \\ | & | \\ -C-C- \\ & | \\ & CN \end{array} \right)$ and 23.4H for water (HOH). In order to produce a polymer solution or lacquer, the paint chemist selects a solvent having a δ value similar to that of the polymer. For example, cellulose dinitrate, which has a δ value of 10.5H, is soluble in Cellosolve ($H_3COCH_2CH_2OH$), which has a δ value of 9.9H.

It is of interest to note that collodion was prepared over a century ago by dissolving cellulose dinitrate in a mixture of ethanol ($H(CH_2)_2OH$) and ethyl ether ($H(CH_2)_2O(CH_2)_2H$). This polymer is not soluble in either of these solvents which have δ values of 12.7H and 7.4H respectively. However, cellulose dinitrate is soluble in a equimolar mixture of these solvents which has a δ value of about 10H.

In contrast to the coatings technologist, the plastics engineer is usually interested in selecting plastics that will not be adversely affected by specific solvents, such as mineral oils, gasoline, ethanol, or water. Accordingly, he selects a plastic with a δ value that differs considerably from the solvent. For example, polytetrafluorethylene, which has a δ value of 6.2H, repels water, which has a δ value of 23.4H. Likewise nylon-66, with a δ value of 13.6H, is not affected adversely by gasoline, with a δ value of about 8H.

Linear polymers or branched polymers such as polystyrene (δ = about 9H) and poly(methyl methacrylate) (δ = about 9.5H) are soluble in appropriate solvents, and, as stated previously, are classified as thermoplastics. However, crosslinked or network polymers, such as phenolic resins or ebonite, will swell in solvents with similar δ values. However, these *thermosetting plastics* will not dissolve in solvents.

1.17 THE EFFECT OF CORROSIVES

Unlike metals, polymers are not affected by the classical electrolytic corrosion processes. Polymers such as poly(vinyl alcohol) are dissolved by aqueous acids and alkalies because these polymers and water have similar δ values. In contrast, polymers with low solubility parameter values are not attacked by water or other polar solvents.

Accordingly, crystalline polymeric hydrocarbons, such as linear polyethylene, are not adversely affected by water. In addition, the carbon-carbon bonds in polymer backbones are not cleaved even at the temperature of boiling water. When amide $\left(\begin{array}{c} O \ H \\ \| \ | \\ -C-N- \end{array}\right)$, ester $\left(\begin{array}{c} O \\ \| \\ -C-O- \end{array}\right)$, or urethane $\left(\begin{array}{c} H \ O \\ | \ \| \\ -N-C-O- \end{array}\right)$ groups are present in the polymer backbone, they may be hydrolyzed by hot aqueous acids or alkalies and hence polymer degradation may occur.

Similar reactions occur when these functional groups are present as pendant groups on the polymer chain. Such attacks reduce polymer integrity but do not reduce the molecular weight appreciably. Nevertheless, the physical properties of a molded plastic will be adversely affected if such an attack occurs. The tendency for such attack is hindered when alkyl groups are present on the carbon atom attached to the functional group or when many stiffening groups, such as phenylene groups, are present in the polymer chain. Thus, poly(methyl methacrylate) and poly(ethylene terephthalate), but not poly(methyl acrylate) or poly(butylene terephthalate), are resistant to acid or alkaline hydrolysis.

Atoms or groups with relatively strong bonds to carbon atoms, such as fluorine (F) and chlorine (Cl), and ether groups are resistant to attack by aqueous acids and alkalies. Thus, polytetrafluoroethylene, poly(vinyl chloride) and chlorinated polyether are seldom attacked by aqueous corrosives. However, the latter is not completely resistant to hot oxidizing acids such as nitric acid, chromic acid or concentrated sulfuric acid.

1.18 AGING OF PLASTICS

Unsaturated polymers, such as natural rubber, are readily attacked by ozone which is present in smog. However, plastics, such as polyethylene, do not contain unsaturated groups and are resistant to attack by ozone. Nevertheless, many plastics are adversely affected by outdoor exposure, even in the absence of ozone.

For example, the energy in the ultraviolet radiations present in our atmosphere will degrade poly(vinyl chloride) in a process called *photodegradation*. It is believed that this is a free radical-type degradation in which an initiating free radical ($R\cdot$) results from absorption of the high energy photons present in ultraviolet light. As shown in Fig. 1-15, this free radical ($R\cdot$) attacks a methylenic hydrogen atom $\left(\begin{array}{c} H \\ | \\ -C- \\ | \\ H \end{array}\right)$ in the polymer chain to produce a macroradical in which a chlorine atom is attached to carbon atoms adjacent to another carbon atom with a single electron. The adjacent atom is called a *beta atom*.

As shown in the equation in Fig. 1-15, the beta chlorine atom leaves as a chlorine atom which is also a free radical, i.e., it is electron deficient since it does not have the full octet of valence electrons which is usually associated with stable molecules. The resulting unsaturated macromolecule is then attacked by the chlorine free radical (Cl·) to produce a polyunsaturated molecule with double bonds on every other carbon atom. This is called a conjugated structure. The polymer becomes dark because of the presence of many unsaturated groups called *polyenes* which are chromophoric or color-producing groups.

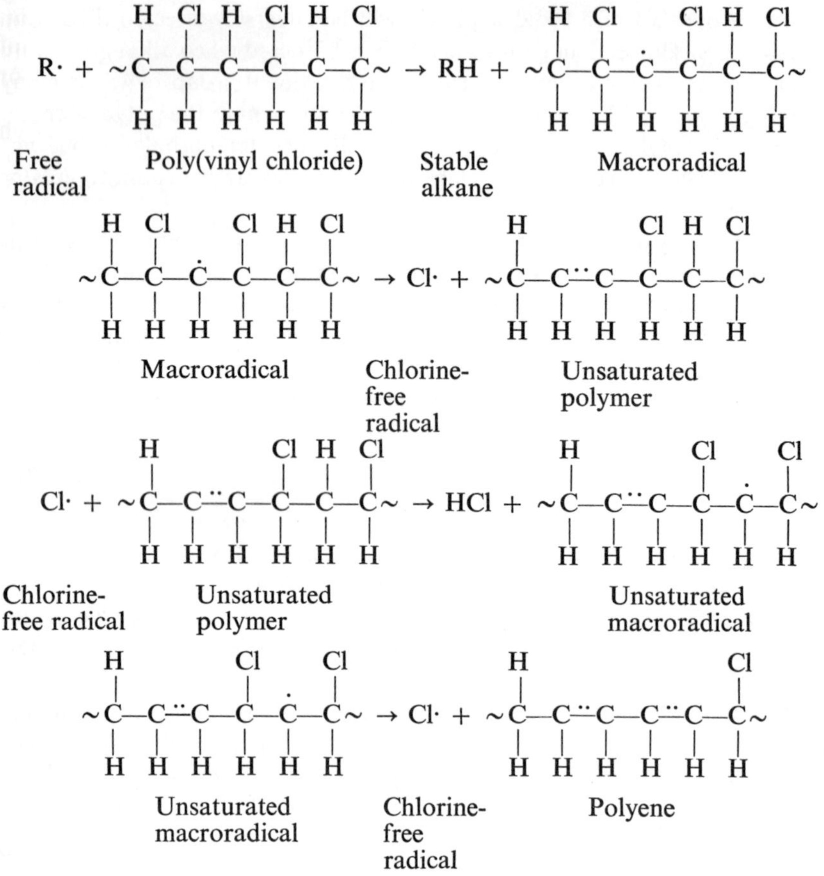

R· + ~CH(Cl)–CH₂–CH(Cl)–CH₂–CH(Cl)–CH₂~ → RH + ~CH(Cl)–CH₂–C·(Cl)–CH₂–CH(Cl)–CH₂~

Free radical / Poly(vinyl chloride) → Stable alkane / Macroradical

~CH(Cl)–CH₂–C·(Cl)–CH₂–CH(Cl)–CH₂~ → Cl· + ~CH=CH–CH₂–CH(Cl)–CH₂~ (shown with dotted C⋯C bond)

Macroradical → Chlorine-free radical / Unsaturated polymer

Cl· + ~CH=CH–CH₂–CH(Cl)–CH₂~ → HCl + ~CH=CH–CH₂–C·(Cl)–CH₂~

Chlorine-free radical / Unsaturated polymer → Unsaturated macroradical

~CH=CH–CH(·)–CH(Cl)–CH₂~ → Cl· + ~CH=CH–CH=CH–CH₂~

Unsaturated macroradical → Chlorine-free radical / Polyene

Fig. 1-15 Proposed mechanism for the photolytic degradation of poly(vinyl chloride) (PVC).

Ultaviolet radiation-initiated degradation is called a *chain reaction* because new free radicals, capable of continuing the reaction, are produced in subsequent steps. Poly(vinylidene chloride) $\left(\begin{array}{c} H\ Cl \\ -C-C- \\ H\ Cl \end{array}\right)$, which has two chlorine atoms on every other carbon atom, also degrades by a similar mechanism.

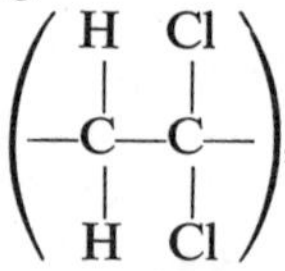

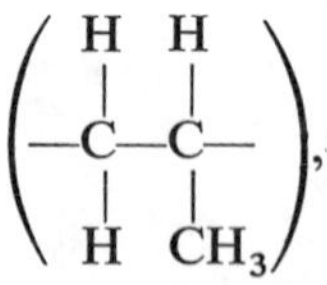

Other polymers, such as polypropylene $\left(\begin{array}{c} H\ H \\ -C-C- \\ H\ CH_3 \end{array}\right)$, are degraded by

ultraviolet light as well as by oxidative degradation. The latter degradative process is also considered to be a free radical chain reaction. As shown in Fig. 1-16, an unstable hydroperoxide may be formed when oxygen reacts with the hydrogen atom on carbon no. 2, called a *tertiary hydrogen atom*. This hydroperoxide then readily dissociates to yield a hydrogen peroxy-free radical (·OOH) and a macroradical. The latter readily loses a methylenic hydrogen atom to produce an unsaturated polymer which continues to degrade in repetitive steps.

These degradative reactions may be retarded by the use of appropriate stabilizers. These additives are discussed in Chap. 2.

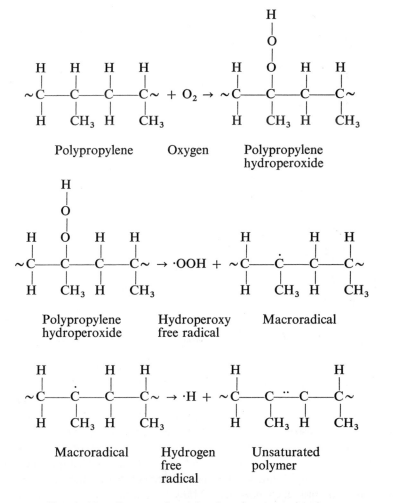

Fig. 1-16 Proposed mechanism for the oxidative degradation polypropylene.

1.19 THE EFFECT OF HEAT

The carbon-carbon bonds in polymer backbones have high bond dissociation energies of at least 80 kcal/mole. Because of insufficient kinetic energy in the molecule, these bonds are not cleaved at moderate temperatures. Thus, organic polymers are usually stable at temperatures as high as 200°C. However, at somewhat higher temperatures, linear polymers are degraded by heat in a process that is often the reverse of polymerization. This depropagation reaction takes place more readily when weak links or imperfections are present in the polymer.

Poly(methyl methacrylate) depolymerizes at elevated temperatures to produce almost quantitative yields of monomeric methyl methacrylate. Bulky pendant groups may also be thermally degraded independently. For example, isobutene ($H_2C{=}C(CH_3)_2$) is obtained when poly(butyl methacrylate) is pyrolyzed. Other polymers, such as polyolefins, decompose into fragments that are larger than the corresponding monomers. The pyrolysis of polystyrene yields both styrene monomer and larger chain fragments. A polymer with a labile chlorine group, such as poly(vinyl chloride) produces hydrochloric acid.

When polyacrylonitrile is heated at temperatures above 220°C, it cyclizes and produces a dark heat-resistant product consisting of fused heterocyclic rings, as shown in Fig. 1-17.

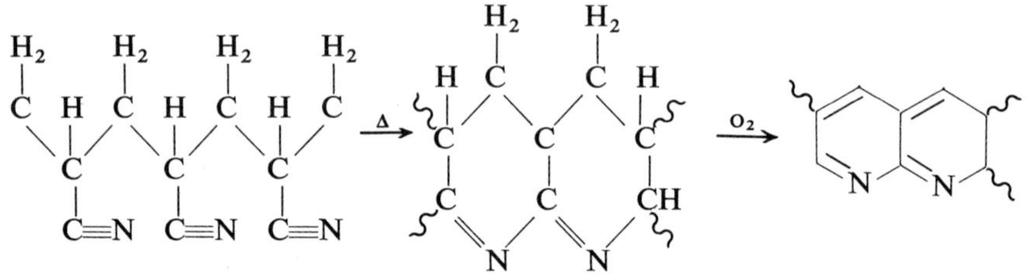

Fig. 1-17 Thermal cyclization of polyacrylonitrile.

A close examination of the structural formula for the PAN pyrolytic endproduct shows that it has a fused laddered structure. In the previously discussed thermal degradation of linear or branched polymers, the cleavage of one bond was required for the production of lower molecular weight products. However, lower molecular weight products cannot be produced from ladder molecules until two distinct bonds are cleaved.

Thus polymers with ladder structures, such as the endproduct of the pyrolysis of polyacrylonitrile, are much more resistant to thermal degradation than linear or branched polymers. Many heat resistant plastics, such as the polyimides, have comparable ladder structures.

BEHAVIORAL OBJECTIVES

Plastics are called macromolecules because they are hundreds of times larger than ordinary molecules. Many of the interesting properties of plastics are dependent on their high molecular weight and the interaction of these long chains.

After reading this chapter, you should understand the following concepts:

1. Polymers or macromolecules are giant molecules with molecular weights at least 100 times those of ordinary molecules such as water or ethanol.

2. Most things we encounter in everyday life such as fibers, rubber, paints, adhesives, starch, protein, nucleic acids and plastics are polymers but in this textbook we will focus our attention on plastics which are the most widely used synthetic polymers.

3. All organic plastics consist of long chains containing sequences of carbon atoms and other atoms such as oxygen, nitrogen, or sulfur. For example, there are usually more than 2,000 carbon atoms joined together in the polyethylene molecule.

4. The macromolecule may assume many different shapes or conformations but it is seldom stretched out to its full contour length. It is customary to use a statistical approach to estimate the average length of macromolecules.

5. Because of increased segmental motion of the polymer chains, the flexibility of a plastic such as polyethylene increases as the temperature is increased.

6. In Isotactic polypropylene, the pendant methyl groups on alternating carbon atoms are all on the same side of the polymer chain. This stereoregular polymer is more rigid and has a higher melting point than an atactic polymer in which the pendant groups are randomly arranged on each side of the polymer chain.

7. Bulky pendant groups increase the volume of a macromolecule and hence decrease its density.

8. Polymers with regular shapes tend to form crystals when cooled. Thus HDPE is opaque because of the presence of aggregates of crystals called spherulites. Non-crystalline plastics are transparent.

9. Random copolymers which contain two or more different building units in the polymer chain such as the copolymer of ethylene and propylene are amorphous and are less rigid than either of the corresponding homopolymers of ethylene or propylene.

10. The temperature at which a stiff plastic becomes flexible but does not melt is called the glass transition temperature (T_g). The melting point temperature (T_m) is usually 33 to 100 percent greater than T_g.

11. The cold flow of linear polymers such as PE may be reduced by the formation of crosslinks between the polymer chains.

12. Fillers restrict the mobility of polymer chains and plasticizers enhance this mobility.

13. Polymer chains of nonpolar molecules such as polyethylene are attracted to each other because of London or dispersion forces based on the formation of temporary dipoles resulting from the transitory distortion of electron clouds in the molecules.

14. Smaller nonpolar molecules are also attacted to each other because of London forces but the strength of the macromolecules is also increased as the result of chain entanglement.

15. Molecules of polar polymers, such as PVC, are attracted to each other because of dipole-dipole interaction between the hydrogen and chlorine atoms in different chains. These forces are greater in polymers such as nylon-66 in which the attraction is between hydrogen and oxygen atoms.

16. Most plastics consist of a mixture of molecules with different molecular weights and are called polydisperse.

17. The ratio of the weight average molecular weight ($\overline{Mw}$) to the number average molecular weight ($\overline{Mn}$) is a measure of the dispersity of polymers. This ration ($\overline{Mw}/\overline{Mn}$) is 1 for monodisperse polymers.

18. The flexibility of polymers is reduced when stiffening groups like carbon-nitrogen bonds are present in the polymer chain.

19. The presence of bulky pendant groups on the polymer chain reduces the Tg and increases the flexibility of atactic polymers.

20. The solubility parameter (δ) is related to the magnitude of the intermolecular forces and hence may be used to select solvents for plastics. A plastic will dissolve in a solvent having a similar δ value.

21. Hydrocarbon plastics such as HDPE are resistant to attack by acids and alkalies but plastics with hydrolyzable groups such as nylon are not resistant to these corrosives.

22. Crosslinked polymers will swell but do not dissolve in solvents with similar δ values.

23. In the absence of stabilizers, some plastics, having labile groups like those present in PVC and PP, are degraded when exposed to high energy radiation such as ultraviolet light.

24. Most organic polymers are stable at temperatures as high as 200°C. However, most general purpose plastics will decompose at temperatures above 350°C.

GLOSSARY

Alternating copolymer: a copolymer in which the building units are arranged in an alternating sequence in the chain.

Amorphous: noncrystalline.

GLOSSARY

Angstrom: 1×10^{-8} cm.
Atactic polymer: one in which the pendant groups are arranged randomly on each side of the polymer chain.
Bakelite: a phenol-formaldehyde polymer (PF).
Block copolymer: a linear copolymer with long sequences of one of the building units in the chain.
Butyl rubber: a copolymer of isobutylene and isoprene.
CED: cohesive energy density, the intermolecular forces between molecules.
Conformation: the shape of the polymer chain which may vary from a coil to a fully stretched chain.
Copolymer: a macromolecule consisting of more than one building unit in the polymer chain.
Delta (δ): solubility parameter.
DP: degree of polymerization or the number of repeating units present in a polymer chain.
$\overline{DP}$: average degree of polymerization.
Elastomers: rubbers.
Filler: a solid additive which restricts the mobility of polymer chains and may also serve as an extender for the polymer.
Flory, Paul: an American scientist who was awarded the Nobel Prize in 1974 in part for his application of statistics to determine the average chain length of molecules.
Glass transition temperature: a characteristic temperature at which amorphous plastics become flexible as a result of the onset of segmental motion of the polymer chains.
Graft copolymer: a branched copolymer in which the branches consist of sequences of one of the building units.
GPC: gel permeation chromatography, a process in which the molecules are separated according to size.
HDPE: high density polyethylene.
Isotactic polymer: one in which the pendant groups are all on the same side of the polymer chain.
LDPE: low density polyethylene.
Linear polymer: one consisting of a continuous chain with a length that is at least 1000 times greater than the thickness of that chain.
London Forces: dispersion forces which attract polymer chains to each other. These are dependent on the formation of temporary dipoles resulting from transitory distortion of electron clouds in the molecules.
Macromolecule: giant or very large molecules.
Melting point: the temperature at which the solid and liquid phases are in equilibrium.
Methylene group: CH_2.
$\overline{Mn}$: number average molecular weight.

Morphology: the study of the shape of polymers.
$\overline{Mw}$: weight average molecular weight.
Natta, Giulio: an Italian scientist who shared the Nobel Prize with Karl Ziegler for the synthesis of isotactic polypropylene.
Organic plastic: one containing carbon atoms in the molecules.
Pendant group: a side chain on a macromolecule such as the methyl groups on alternating carbon atoms in polypropylene.
PF: phenol-formaldehyde resin.
Plastic: solid high molecular weight materials which are usually processed in a liquid state in order to produce a finished article.
Plasticizer: a liquid or solid additive which increases the mobility of polymer chains.
PMMA: poly(methyl methacrylate).
PP: polypropylene.
PVC: poly(vinyl chloride).
Skeletal formula: a structural formula that does not include the hydrogen atoms.
Solubility parameter: a scale of values related to intermolecular forces.
Spherulites: aggregates of crystals.
Staudinger, Hermann: a German scientist who was awarded the Nobel Prize in 1953 for his explanation of polymer structure.
Syndiotactic polymer: one in which the pendant groups are arranged in an alternating configuration on each side of the polymer chain.
Thermoplastic: a linear or branched fusible polymer.
Thermoset: an infusible crosslinked polymer.

REVIEW QUESTIONS

1. Name five polymers that you encounter daily.
2. Compare the molecular weight of polyethylene with that of water. The molecular weight of water (H_2O) is 18 g.
3. If there are 2000 repeating methylene (CH_2) units in a polyethylene chain and each of these is 1.5Å, what would be the average chain length?
4. Which will have the higher melting point, isotactic polypropylene or atactic polypropylene?
5. Which will have the greater volume—HDPE or LDPE.
6. Which would have the greater tendency to form crystals—LDPE or HDPE?
7. Which would be more flexible, a random copolymer of ethylene and propylene, PP or HDPE?
8. Using A and B for the building units, write the structures for random and block copolymers.

ANSWERS

9. Which would have the greater tendency to cold flow at room temperature, polypropylene (Tg = −18°C) or polyisobutylene (Tg = −70°C)?
10. Which would be more stable at high temperatures, linear polyethylene or crosslinked polyethylene?
11. Which would increase the Tg of a polymer such as PVC more, the addition of asbestos or dioctyl phthalate?
12. What is the principal attractive force between nonpolar polymer molecules such as PE?
13. Why is polyethylene stronger than paraffin wax?
14. Which polymer would have the stronger intermolecular forces, polyethylene or nylon?
15. How do casein and HDPE differ in molecular weight distribution?
16. Why is the nylon chain stiffer than HDPE?
17. Which will have the higher Tg value, poly(methyl methacrylate) or poly(n-butyl methacrylate)?
18. Which size molecules will be eluted first in GPC?
19. Which will have the higher solubility parameter value, cellulose or polyethylene?
20. From a choice of HDPE, PMMA and nylon-66, which plastic would you choose for service in 20% hydrochloric acid at 50°C?
21. What solvent would you select to dissolve melamine plastic?
22. Which plastic is more resistant to ultraviolet light, PVC or HDPE?
23. Why is pyrolized polyacrylonitrile heat resistant?
24. How could you recover methyl methacrylate monomer from PMMA?

ANSWERS

1. This list could include paper, wood, nylon, polyester fibers, melamine dishes, polyethylene film, polytetrafluoroethylene (Teflon) coated cooking ware, potatoes, wool, hair, polypropylene carpets, poly(vinyl acetate) coated walls, etc.
2. The molecular weight of polyethylene is at least 100 times greater than that of water and could be 1,000,000 times greater.
3. Much less than 2,000 (1.5Å) and actually about 6 percent of this 3,000Å full contour length value.
4. Isotactic polypropylene.
5. LDPE.
6. HDPE.
7. The random polymer.
8. —AABABBBAAB—, —AAAAABBBBB—
9. Polyisobutylene.
10. Crosslinked polyethylene.

11. Asbestos.
12. London or dispersion forces.
13. More London forces plus chain entanglement. The latter is not characteristic of smaller molecules such as paraffin wax.
14. Nylon in which both hydrogen bonds and London forces exist between polymer chains.
15. Casein is monodisperse, i.e., all the molecules have the same molecular weight, whereas HDPE is polydisperse and consists of a mixture of polymers of different molecular weights.
16. Because of the presence of stiffening groups in the polymer chain.
17. Poly(methyl methacrylate).
18. The largest, i.e., highest molecular weight.
19. Cellulose since it is more polar.
20. HDPE preferred but PMMA is serviceable.
21. Melamine plastic is a thermoset and hence may swell but will not dissolve in solvents.
22. HDPE.
23. It is a ladder polymer.
24. By thermal decomposition and distillation of the product.

chapter 2

Compounding Ingredients, Initiators, Chain Transfer Agents, and Fillers

2.1 THE USEFULNESS OF ADDITIVES

A clear plastic that is useful for glazing, ornamental uses, and several other applications may be prepared by the thermal polymerization of methyl methacrylate in an appropriately shaped container in the absence of oxygen and without any other additives. However, uncompounded castings, such as cast poly(methyl methacrylate) (PMMA) or epoxy resins (EP) have limited use and most commercial plastics are actually composites consisting of the polymeric binder and appropriately selected compounding ingredients.

Most uncompounded polymers, such as natural rubber (NR), phenolic resins (PF), poly(vinyl chloride) (PVC), and polypropylene (PP) have very limited use. Carbon black is added to rubber, wood-flour filler is added to phenolic resins, dioctyl phthalate plasticizer is added to PVC and a hindered phenol antioxidant is added to PP so that these polymers may have optimum usefulness.

2.2 INITIATORS

Most commercial synthetic polymers are produced by a chain reaction polymerization process, sometimes called *addition polymerization*. The fission of uranium or plutonium, which produces energy, the combustion of fossil fuels and the addition polymerization are all chain reactions requiring an initiator to start the self-sustaining reactions. Polymerization initiators may be inorganic protonic acids, such as sulfuric acid (H_2SO_4), Lewis acids, such as aluminum chloride ($AlCl_3$), organometallic compounds, such as butyllithium (C_4H_9Li), a combination of an organometallic compound, such as triethylaluminum ($Al(C_2H_5)_3$) and a salt of a transition metal, such as titanium (IV) chloride ($TiCl_4$), called the *Ziegler-Natta Catalyst*, or unstable compounds which are readily decomposed. The latter group consists of molecules with weak bonds which are easily cleaved to produce active products called *free radicals*. Typical examples of these initators are inorganic peroxycompounds, such as potassium persulfate ($K_2S_2O_8$), organic peroxycompounds, such as benzoyl peroxide (($C_6H_5CO_2)_2$) and organic azo compounds, such as azobisisobutyronitrile (AIBN)(($CH_3)_2C(CN)N)_2$).

Relatively small amounts of Lewis acids, such as aluminum chloride ($AlCl_3$), are used to initiate the polymerization of unsaturated hydrocarbons with electropositive substituents, such as isobutylene ($H_2C:C(CH_3)_2$), for the production of polyisobutylene. Since the true initiator is actually a proton (H^+) and the growing chain is a macrocation, this process is called *cationic polymerization*. As shown in Fig. 2-1, the proton is believed to be produced by the dissociation of the reaction product of aluminum chloride ($AlCl_3$)

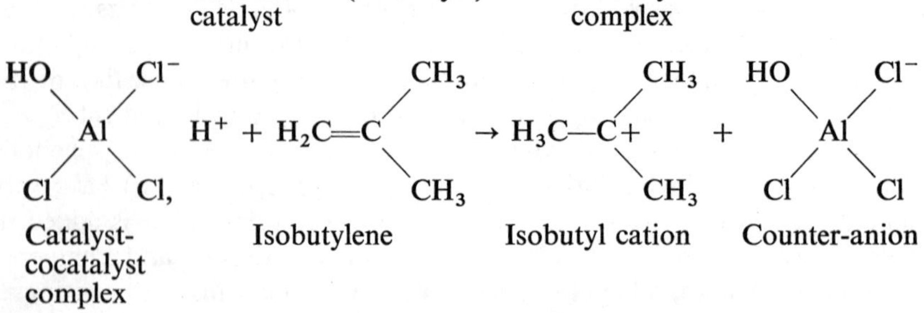

Fig. 2-1 Cationic initiation of isobutylene.

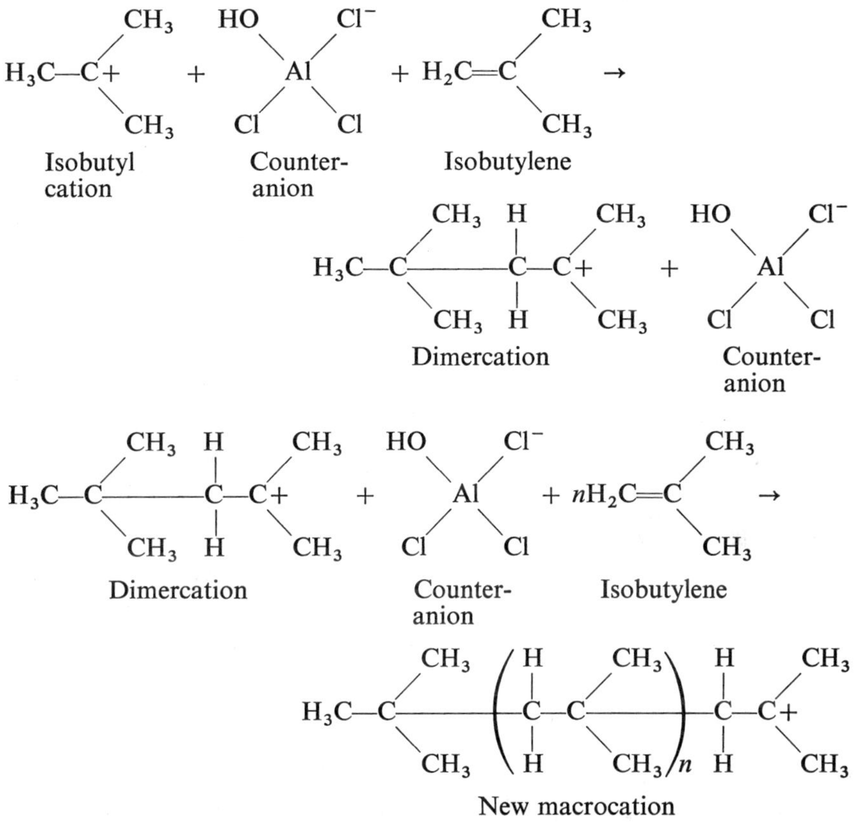

Sec. 2.2 INITIATORS

$$H_3C-\underset{CH_3}{\overset{CH_3}{C}}+ \;+\; \underset{Cl\quad Cl}{\overset{HO\quad Cl^-}{Al}} \;+\; H_2C=\underset{CH_3}{\overset{CH_3}{C}} \;\rightarrow$$

Isobutyl cation   Counteranion   Isobutylene

$$H_3C-\underset{CH_3}{\overset{CH_3}{\underset{|}{C}}}-\underset{H}{\overset{H}{\underset{|}{C}}}-\underset{CH_3}{\overset{CH_3}{C}}+ \;+\; \underset{Cl\quad Cl}{\overset{HO\quad Cl^-}{Al}}$$

Dimercation   Counteranion

$$H_3C-\underset{CH_3}{\overset{CH_3}{\underset{|}{C}}}-\underset{H}{\overset{H}{\underset{|}{C}}}-\underset{CH_3}{\overset{CH_3}{C}}+ \;+\; \underset{Cl\quad Cl}{\overset{HO\quad Cl^-}{Al}} \;+\; nH_2C=\underset{CH_3}{\overset{CH_3}{C}} \;\rightarrow$$

Dimercation   Counteranion   Isobutylene

$$H_3C-\underset{CH_3}{\overset{CH_3}{C}}\left(-\underset{H}{\overset{H}{\underset{|}{C}}}-\underset{CH_3}{\overset{CH_3}{C}}-\right)_n \underset{H}{\overset{H}{\underset{|}{C}}}-\underset{CH_3}{\overset{CH_3}{C}}+$$

New macrocation

Fig. 2-2 Propagation of isobutylene.

and water (H_2O). Accordingly, water is called a *cocatalyst*. The reaction product of aluminum chloride and water is called a *catalyst-cocatalyst* complex.

The isobutyl cation, like the proton, is attracted to an isobutylene molecule to form a dimercation. This type of reaction is repeated in many subsequent propagation reactions, as shown in Fig. 2-2. The product of each propagation step is a larger macrocation.

The propagation of these macrocations may be terminated by chain transfer of a proton from the macrocation to a counteranion to produce a dead polymer with an unsaturated chain end and a catalyst-cocatalyst complex, or a dead saturated polymer when a hydide ion ($H:^-$) is transferred from a compound like toluene as illustrated in Fig. 2-3. The new cationic species produced in both chain-transfer reactions are capable of initiating new polymerization reactions.

Relatively small amounts of an organometallic compound, such as butyllithium (C_4H_9Li), are used to initiate the polymerization of unsaturated

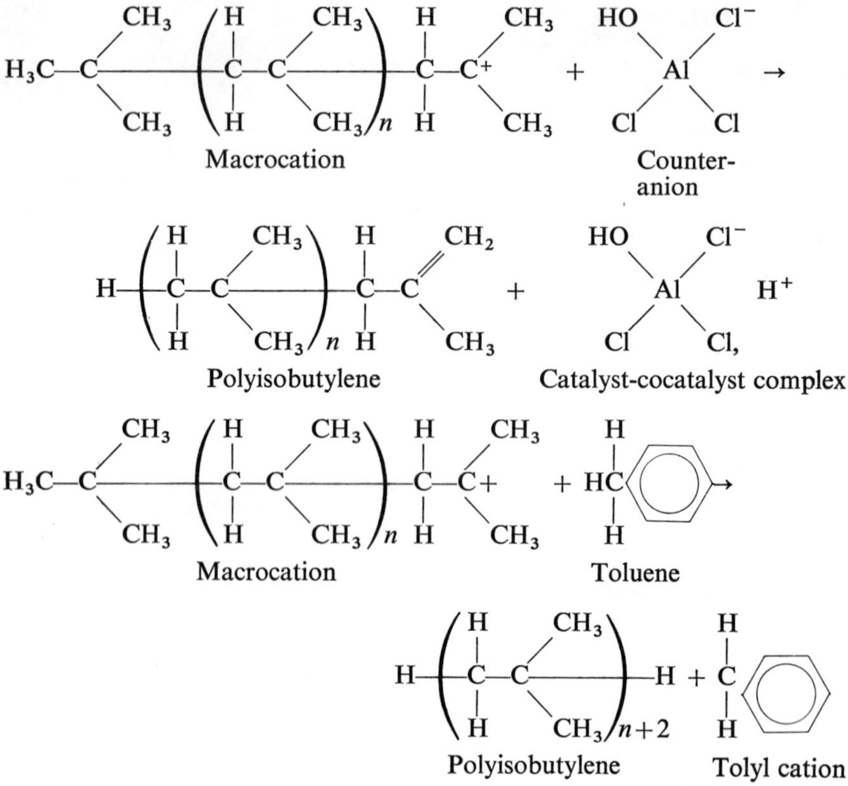

Fig. 2-3 Termination of propagation of isobutylene by chain transfer.

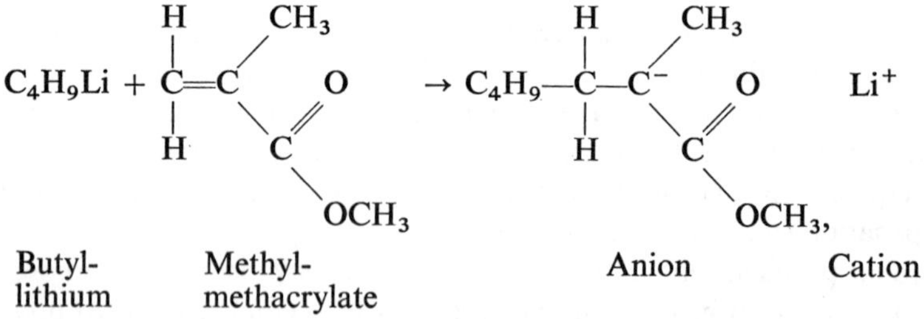

Fig. 2-4 Anionic initation of methyl methacrylate.

monomers with electronegative substituents, such as methyl methacrylate ($H_2C:C(CH_3)COOCH_3$). In contrast with the previously discussed cationic propagation, the propagating species formed by initiation by a base such as butyllithium is an anion. The initiation and propagation steps for anionic polymerization are shown in Figs. 2-4 and 2-5.

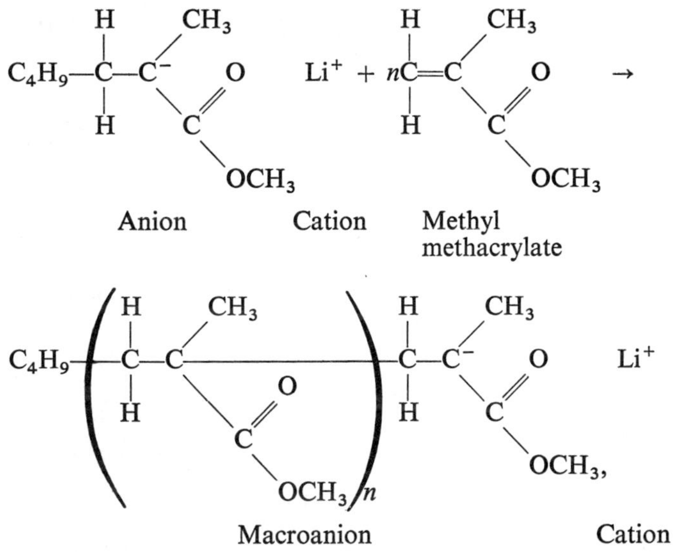

Sec. 2.2 INITIATORS

$$C_4H_9-\underset{H}{\overset{H}{C}}-\underset{C(OCH_3)=O}{\overset{CH_3}{C^-}} \quad Li^+ + n\underset{H}{\overset{H}{C}}=\underset{C(OCH_3)=O}{\overset{CH_3}{C}} \rightarrow$$

Anion   Cation   Methyl methacrylate

$$C_4H_9-\left(\underset{H}{\overset{H}{C}}-\underset{C(OCH_3)=O}{\overset{CH_3}{C}}\right)_n -\underset{H}{\overset{H}{C}}-\underset{C(OCH_3)=O}{\overset{CH_3}{C^-}}, \quad Li^+$$

Macroanion   Cation

Fig. 2-5 Anionic propagation of methyl methacrylate.

Unsaturated hydrocarbons, such as propylene ($H_2C:C(CH_3)H$), are polymerized by Ziegler-Natta catalysts in a coordination polymerization process. While the mechanism for coordination polymerization is not completely understood, it is assumed that the initator is a complex formed from the titanium tetrachloride and aluminum triethyl, as shown in Fig. 2-6. It is also assumed that propagation proceeds by an insertion mechanism involving a breaking of the carbon-metal bonds, illustrated in Fig. 2-6. These reactions are terminated by the addition of hydrogen or hydrogen compounds such as ethanol.

The most widely used initiators are free radicals produced by the dissociation of unstable peroxides or azo compounds. The covalent bonds in these initiators are much weaker than those in stable compounds. In contrast to to the normal bond dissociation energy of the carbon-carbon bond, which is about 90 kcal/mole in most organic compounds, the bond dissociation energy of the oxygen-oxygen atoms in hydrogen peroxide is 51 kcal/mole. Thus, hydrogen peroxide readily forms hydroxyl-free radicals ($\cdot OH$) in an aqueous system such as that used in the emulsion polymerization process. Potassium persulfate ($K_2S_2O_8$) also dissociates readily in aqueous solutions to produce sulfate-free radical ions ($\cdot SO_4^-$).

Polymers such as PVC and poly(vinyl acetate) (PVAC) are usually produced by the emulsion polymerization technique, as shown in Fig. 2-7, in which the initiation step takes place in the aqueous phase. Fig. 2-8 shows these free radical ions adding to vinyl monomers, such as vinyl acetate ($H_2C:C(H)OOCCH_3$), to produce new free radical ions.

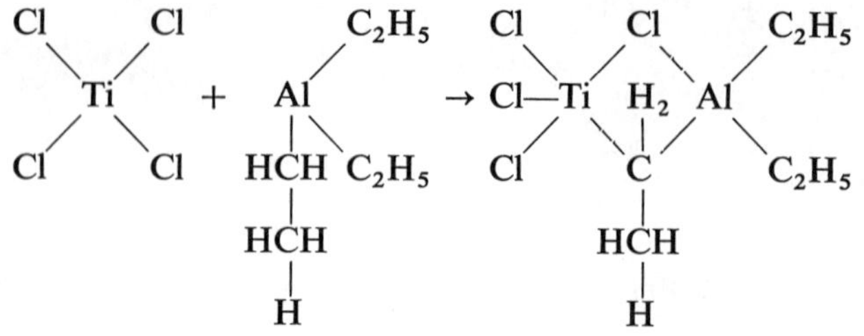

$$\underset{\substack{\text{Titanium}\\\text{tetrachloride}}}{\begin{array}{c}\text{Cl}\quad\text{Cl}\\\diagdown\diagup\\\text{Ti}\\\diagup\diagdown\\\text{Cl}\quad\text{Cl}\end{array}} + \underset{\substack{\text{Aluminum-}\\\text{triethyl}}}{\begin{array}{c}\quad\ \text{C}_2\text{H}_5\\\diagup\\\text{Al}\\\diagdown\\\text{HCH}\ \text{C}_2\text{H}_5\\|\\\text{HCH}\\|\\\text{H}\end{array}} \rightarrow \underset{\text{Coordination complex}}{\begin{array}{c}\text{Cl}\qquad\qquad\ \ \text{C}_2\text{H}_5\\\diagdown\qquad\qquad\diagup\\\text{Cl—Ti}\ \text{H}_2\ \text{Al}\\\diagup\ \ \diagdown\diagup\ \diagdown\\\text{Cl}\qquad\text{C}\qquad\text{C}_2\text{H}_5\\|\\\text{HCH}\\|\\\text{H}\end{array}}$$

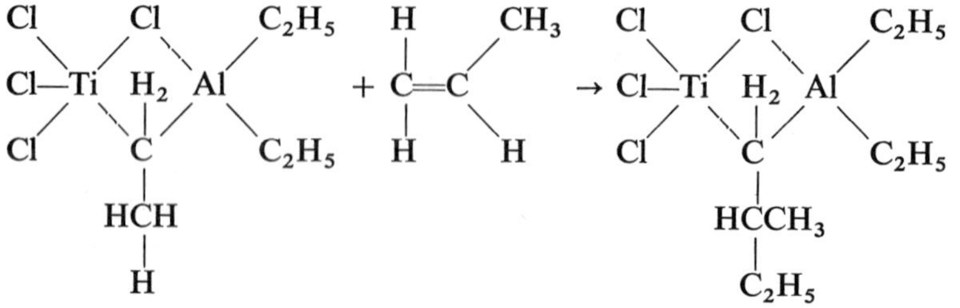

$$\underset{\text{Coordination complex}}{\begin{array}{c}\text{Cl}\qquad\qquad\ \ \text{C}_2\text{H}_5\\\diagdown\qquad\qquad\diagup\\\text{Cl—Ti}\ \text{H}_2\ \text{Al}\\\diagup\ \ \diagdown\diagup\ \diagdown\\\text{Cl}\qquad\text{C}\qquad\text{C}_2\text{H}_5\\|\\\text{HCH}\\|\\\text{H}\end{array}} + \underset{\text{Propylene}}{\begin{array}{c}\text{H}\quad\text{CH}_3\\\diagdown\diagup\\\text{C}=\text{C}\\\diagup\diagdown\\\text{H}\quad\text{H}\end{array}} \rightarrow \underset{\text{New coordination complex}}{\begin{array}{c}\text{Cl}\qquad\qquad\ \ \text{C}_2\text{H}_5\\\diagdown\qquad\qquad\diagup\\\text{Cl—Ti}\ \text{H}_2\ \text{Al}\\\diagup\ \ \diagdown\diagup\ \diagdown\\\text{Cl}\qquad\text{C}\qquad\text{C}_2\text{H}_5\\|\\\text{HCCH}_3\\|\\\text{C}_2\text{H}_5\end{array}}$$

Fig. 2-6 Suggested mechanism for Ziegler-Natta polymerization.

$$\underset{\substack{\text{Persulfate}\\\text{ion}}}{\text{S}_2\text{O}_8^=} \rightarrow \underset{\substack{\text{Sulfate-free}\\\text{radical ion}}}{2\text{SO}_4^-.}$$

Fig. 2-7 Dissociation of persulfate ions to produce free radicals.

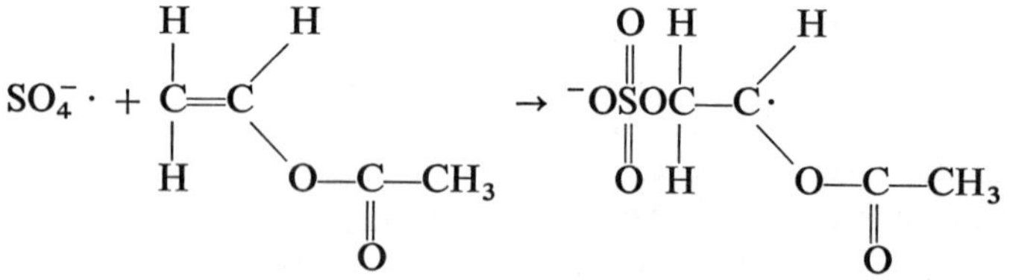

$$\underset{\substack{\text{Sulfate}\\\text{free}\\\text{radical}\\\text{ion}}}{\text{SO}_4^-\cdot} + \underset{\substack{\text{Vinyl}\\\text{acetate}}}{\begin{array}{c}\text{H}\quad\ \ \text{H}\\|\qquad\diagup\\\text{C}=\text{C}\\|\qquad\diagdown\\\text{H}\quad\ \ \text{O—C—CH}_3\\\qquad\qquad\|\\\qquad\qquad\text{O}\end{array}} \rightarrow \underset{\substack{\text{New free radical}\\\text{ion}}}{\begin{array}{c}\quad\ \text{O}\ \text{H}\quad\ \ \text{H}\\\quad\ \|\ |\qquad\diagup\\{}^-\text{OSOC—C}\cdot\\\quad\ \|\ |\qquad\diagdown\\\quad\ \text{O}\ \text{H}\quad\ \ \text{O—C—CH}_3\\\qquad\qquad\qquad\ \|\\\qquad\qquad\qquad\ \text{O}\end{array}}$$

Fig. 2-8 Initiation of vinyl polymerization by free radicals.

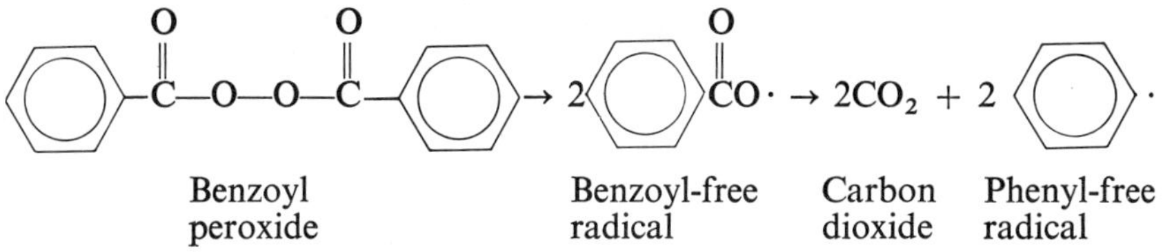

Sec. 2.2 INITIATORS 39

Benzoyl peroxide → 2 **Benzoyl-free radical** → $2CO_2$ + 2 **Phenyl-free radical**

(Carbon dioxide)

Fig. 2-9 Decomposition of benzoyl peroxide.

These new free radicals migrate to aggregates of soap molecules called *micelles*. Propagation takes place within these spherical soap aggregates. The hydrocarbon end ($H(CH_2)_{17}^-$) of the soap molecule is oriented away from the water phase and the carboxylic end ($-COO^-$) is oriented toward the water phase. The rate of termination resulting from the coupling of two free radicals is retarded because the macroradical is protected by the micelle.

Organic peroxides, such as benzoyl peroxide, also dissociate readily to produce benzoyl free radicals which in turn decompose to form carbon dioxide (CO_2) and phenyl free radicals ($\cdot C_6H_5$), as depicted in Fig. 2-9.

The phenyl free radicals then add to a vinyl monomer, such as styrene ($C_6H_5CH:CH_2$), to produce a new free radical which can propagate with other vinyl monomers to form macroradicals, as shown in Fig. 2-10.

Termination of this chain propagation may take place by the coupling of two free radicals or by the abstraction of a hydrogen atom from one macroradical by another macroradical in a process called disproportionation. Typical termination reactions are illustrated in Fig. 2-11.

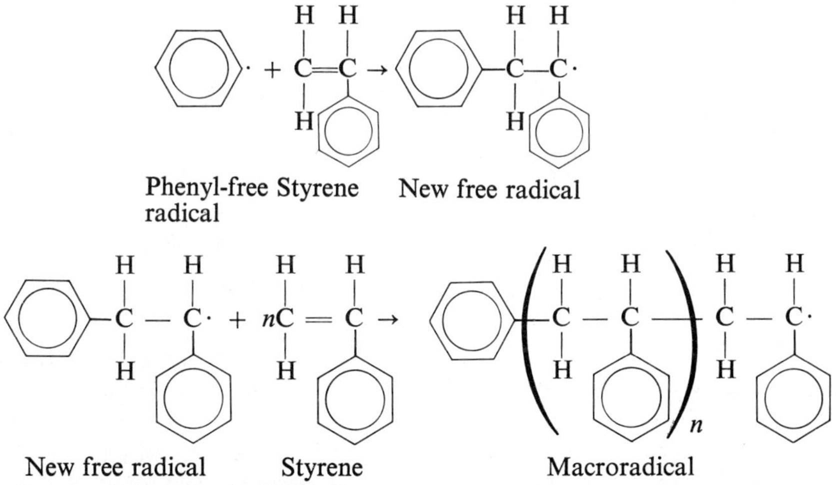

Fig. 2-10 Free radical initiation and propagation of styrene.

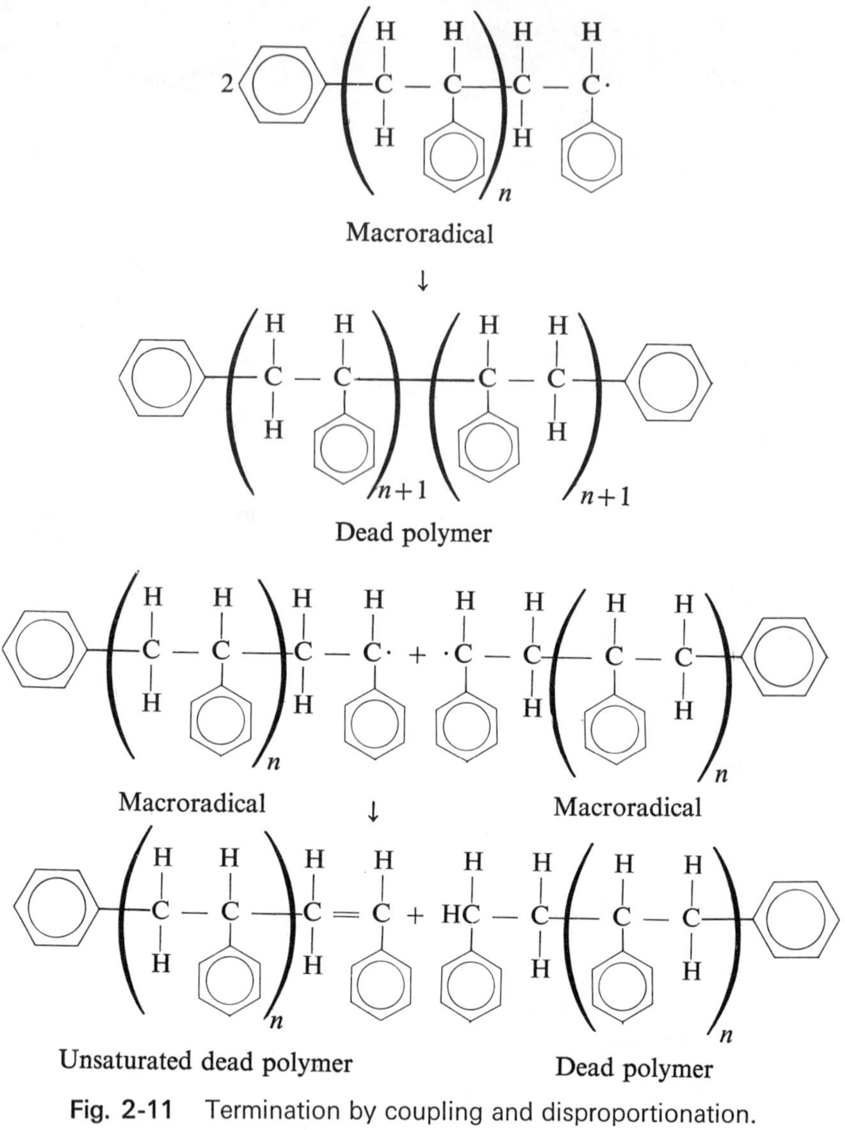

Fig. 2-11 Termination by coupling and disproportionation.

The most widely used organic initiator, benzoyl peroxide, is often supplied as a slurry or paste consisting of equal parts of the initiator and a liquid such as dibutyl phthalate. Since organic peroxides are readily decomposed, they must be handled with caution. Their decomposition may be accelerated by heat, by the addition of small amounts of tertiary amines, such as N,N-dimethylaniline $(C_6H_5N(CH_3)_2)$ or by organic salts of heavy metals such as cobalt naphthenate.

TABLE 2-1 The Half Lives of Common Organic Peroxides at 50°C

Organic Peroxide	Half Life $(T_{1/2})$ (hrs)
Lauroyl peroxide	54
Acetyl peroxide	158
Benzoyl peroxide	190
Methyl ethyl ketone peroxide	10,000
Tert. butyl hydroperoxide	100,000
Cumene hydroperoxide	500,000

The relative stability of organic peroxides is usually expressed in terms of half lives. The rate of decomposition at any specific temperature is proportional to the concentration of the initiator and the time for 50% of the sample to decompose is its half life. The half lives of some of the more common organic peroxides are shown in Table 2-1.

As shown in Fig. 2-12, azo compounds such as azobisisobutyronitrile (AIBN) decompose to produce nitrogen and free radicals when heated or exposed to ultraviolet radiation.

$$(H_3C)_2-\underset{CN}{C}-N=N-\underset{CN}{C}-(CH_3)_2 \rightarrow N_2 + 2(H_3C)_2-\underset{CN}{C}\cdot$$

Azobisisobutyronitrile   Nitrogen   Free radical

Fig. 2-12 The decomposition of azobisisobutyronitrile.

It is important to note that initiators, such as benzoyl peroxide, potassium persulfate and AIBN are not true catalysts since the free radical group is present on the chain end of the dead polymer. (See Fig. 2-10.)

2.3 CHAIN TRANSFER AGENTS, MODIFIERS, PEPTIZERS

Low molecular weight polymers called *oligomers* are not useful as plastics because the chains are not long enough to permit entanglement. Polymers with moderate chain lengths may be used as protective coatings but polymer chains consisting of less than 100 mers are not satisfactory for use as plastics. In contrast, the extremely high molecular weight polymers that are produced in emulsion polymerization are often unsatisfactory because of the difficulties associated with the mechanical processing of extremely long chained molecules. While these supermacromolecules do possess improved abrasion resistance, the added costs of processing is seldom justified for routine plastics applications. Thus, it is customary to add controlled amounts of foreign materials, which act as transfer agents, and produce polymers with moderate molecular weights.

As discussed earlier in Sec. 2.2, cationic propagation may be terminated by the transfer of a hydride ion from toluene. Likewise, free radical propagation may be terminated by the transfer of a loosely bonded atom such as hydrogen from a chain transfer agent, such as carbon tetrachloride (CCl_4) or dodecyl mercaptan ($H(CH_2)_{12}SH$). As shown in Fig. 2-13, this additive, also called a *modifier*, loses a hydrogen atom by homolytic cleavage of the covalent bond and the propagating macroradical ($\sim P\cdot$) is terminated by coupling with this atom. The products of this chain transfer are a dead polymer and a new free radical.

$$\sim P\cdot\ +\ HS(CH_2)_{12}H\ \rightarrow\ PH\ +\ \cdot S(CH_2)_{12}H$$

| Macro-radical | Dodecyl mercaptan | Dead polymer | New free radical |

Fig. 2-13 A typical chain transfer reaction.

The chain transfer constant, which is a measure of chain transfer efficiency, is determined experimentally by keeping all other polymerization variables constant while adding increasing amounts of a chain transfer agent or telogen to a polymerization system and determining the average molecular weight of the polymeric product. The chain transfer constant (C_S) is the slope of the line obtained when the reciprocal of the chain length ($\overline{DP}^{-1}$) is plotted against the ratio of the concentrations of the telogen and the monomer. The low molecular weight product obtained in the presence of high concentrations of the telogens is called a *telomer* and the process is called *telomerization*.

Since the new free radicals initate new polymer chains which may be terminated by chain transfer, most of the dead polymer chains contain fragments of the telogen molecule as chain ends. Thus, when carbon tetrachloride or carbon tetrabromide are used as telogens, the resulting telomers contain terminal halogen atoms. Other functional groups, such as carboxyl (—COOH), hydroxyl (—OH), and amino groups (—NH_2), are present in telomers produced in the presence of organic acids, aliphatic alcohols, and amines respectively.

It is customary to add a small amount of an inhibitor such as tert. butylcatechol to prevent the premature or accidental polymerization of monomers such as styrene. If this additive is not removed before polymerization, it will act as a chain transfer agent and inhibitor fragments may be present in the polymeric product.

Many commercial polymers, such as polyamides (nylon) (PA), polyesters, polyurethanes (PUR), polyethers and epoxy (EP), phenolic (PF), urea (UF), and melamine resins (MF), are produced by a step reaction or condensation process in which polymer chains are built up simultaneously in a step reaction process. This process differs from the previously discussed chain

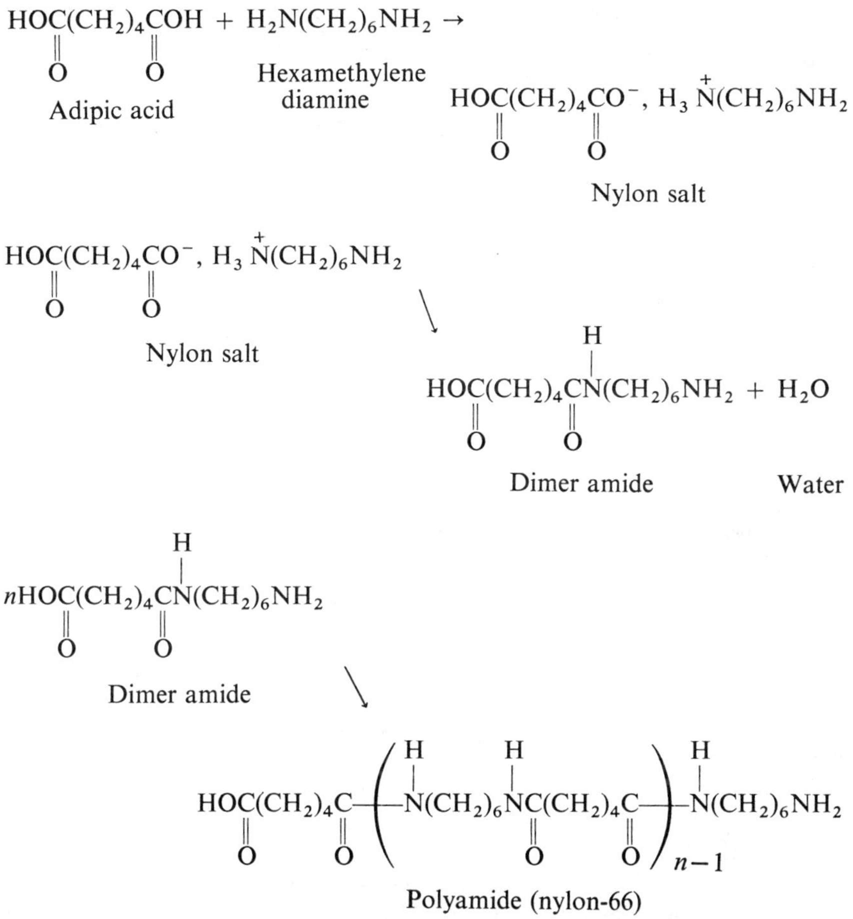

Sec. 2.3 CHAIN TRANSFER AGENTS, MODIFIERS, PEPTIZERS

$$\text{HOC(CH}_2)_4\text{COH} + \text{H}_2\text{N(CH}_2)_6\text{NH}_2 \rightarrow$$
$$\|\|$$
$$\text{O}\text{O}$$

Adipic acid   Hexamethylene diamine

$$\text{HOC(CH}_2)_4\text{CO}^-, \text{H}_3\overset{+}{\text{N}}(\text{CH}_2)_6\text{NH}_2$$
$$\|\|$$
$$\text{O}\text{O}$$

Nylon salt

$$\text{HOC(CH}_2)_4\text{CO}^-, \text{H}_3\overset{+}{\text{N}}(\text{CH}_2)_6\text{NH}_2$$

Nylon salt

$$\searrow$$

$$\text{HOC(CH}_2)_4\overset{}{\underset{\|}{\text{C}}}\overset{\text{H}}{\underset{}{\text{N}}}(\text{CH}_2)_6\text{NH}_2 + \text{H}_2\text{O}$$

Dimer amide   Water

$$n\text{HOC(CH}_2)_4\overset{\text{H}}{\underset{\|}{\text{C}}}\overset{}{\underset{}{\text{N}}}(\text{CH}_2)_6\text{NH}_2$$

Dimer amide

$$\searrow$$

$$\text{HOC(CH}_2)_4\text{C}\!\left(\!-\overset{\text{H}}{\text{N}}(\text{CH}_2)_6\overset{\text{H}}{\text{N}}\text{C}(\text{CH}_2)_4\text{C}\!-\!\right)_{\!n-1}\!\!-\overset{\text{H}}{\text{N}}(\text{CH}_2)_6\text{NH}_2$$

Polyamide (nylon-66)

Fig. 2-14 The preparation of nylon-66 from adipic acid and hexamethylenediamine.

reaction polymerization in which the macroion or macroradical continued to add monomer units until termination by coupling, disproportionation, or chain transfer.

The process resembles simple organic chemical reactions between monofunctional groups but the products produced in the step reactions of molecules with bifunctional groups contain residual functional groups that are capable of further reaction. For example, as shown in Fig. 2-14, nylon-66 is produced by heating the salt produced from the reaction of adipic acid and hexamethylenediamine. The initial product contains carboxylic and amino end groups that are capable of condensing to produce longer chains that, in the absence of foreign materials, will continue to condense to produce still longer chains.

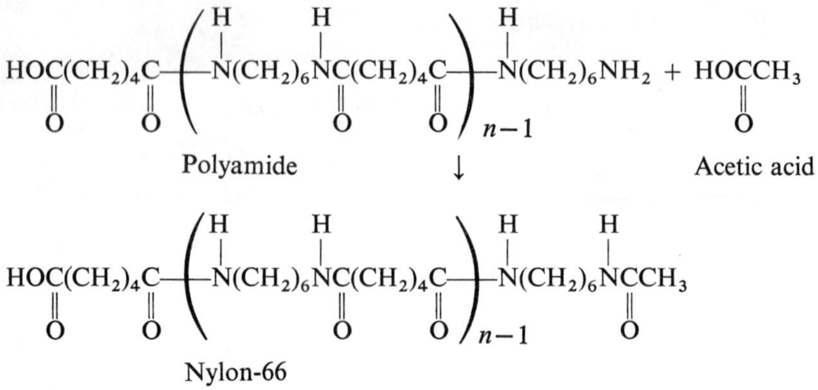

$$\text{HOC(CH}_2)_4\text{C}\underset{\text{O}}{\overset{\text{O}}{\|}}\left(\text{N(CH}_2)_6\overset{\text{H}}{\underset{|}{\text{N}}}\text{C(CH}_2)_4\underset{\text{O}}{\overset{\text{O}}{\|}}\right)_{n-1}\overset{\text{H}}{\underset{|}{\text{N}}}(\text{CH}_2)_6\text{NH}_2 + \text{HOCCH}_3\underset{\text{O}}{\overset{\text{O}}{\|}}$$

Polyamide $\quad\downarrow\quad$ Acetic acid

$$\text{HOC(CH}_2)_4\text{C}\left(\text{N(CH}_2)_6\text{NC(CH}_2)_4\text{C}\right)_{n-1}\text{N(CH}_2)_6\text{NCCH}_3$$

Nylon-66

Fig. 2-15 Termination of a growing nylon-66 chain by condensation with acetic acid.

Since extremely high molecular weight molecules are difficult to process, a small amount of a monofunctional reactant such as acetic acid is added to the nylon salt before thermal dehydration. Figure 2-15 shows that a dead polymer will be produced whenever acetic acid reacts with the amino group in a growing chain.

The amount of acetic acid required to produce a chain of any average length may be calculated from a modification of Carothers' equation, which states that the degree of polymerization is inversely proportional to $1 - P$ where the extent of the polymerization reaction is represented by P. The effect of the addition of any mole fraction of acetic acid (na/n) on the average degree of polymerization ($\overline{DP}$) is depicted in Fig. 2-16.

The effect of the addition of 1 mole of acetic acid to a reaction mixture containing 99 moles of adipic acid may be calculated as follows for a condensation where the extent of polymerization (P) is 0.999. In the absence of the acetic acid, the average degree of polymerization $\overline{DP}$ would be $1/(1 - 0.999) = 1000$. When the mole fraction of acetic acid present is 0.01, the $\overline{DP}$ is reduced to $(1 + 0.01)/(1 - 0.999 + 0.01) = 101$.

When the average molecular weight is so high as to make mechanical processing difficult, chain transfer agents, called *peptizers*, may be added to reduce the average molecular weight. Mechanical processing, such as mastication on a two roll mill, causes a cleavage of the covalent bonds in

$$\overline{DP} = \frac{1 + \dfrac{na}{n}}{1 - p + \dfrac{na}{n}}$$

Fig. 2-16 Carothers' equation for na/n mole fraction of monofunctional additive.

Sec. 2.4 REACTIVE ADDITIVES (CURING AGENTS)

$$P{:}P \rightarrow 2P\cdot + 2HR \rightarrow 2PH + 2R\cdot$$
Polymer   Macroradical   Peptizer   Dead polymer   Free radical

Fig. 2-17 Cleavage of polymers in the presence of a peptizer.

polymer chains. Figure 2-17 illustrates that the products of this homolytic cleavage are macroradicals that in the absence of oxygen readily recombine to produce high molecular weight products. However, if peptizers are present, hydrogen atoms or other loosely-bonded atoms are abstracted by the macroradicals and dead polymers are produced.

2.4 REACTIVE ADDITIVES (CURING AGENTS)

The usefulness of natural rubber and some synthetic plastics, such as unsaturated polyester, epoxy (EP) and phenolic resins (PF), is limited unless these linear polymer chains are crosslinked or cured. The modern rubber industry is dependent on the serendipitous discovery of sulfur-curing or vulcanization by Charles Goodyear. In addition, the heat resistance of thermosetting resins, such as phenolic resins is dependent on the addition of a different crosslinking agent.

Linear or *A* stage novolak phenolic resins are produced at an annual rate of over 600 tons by condensing 1 mole of phenol (C_6H_5OH) with 0.84 moles of formaldehyde (H_2CO) in the presence of 0.3g sulfuric acid. It is customary to supply the formaldehyde as a 37% aqueous solution called formalin. The excess water is removed by vacuum distillation after hot condensation of the phenol and formaldehyde, and the molten resin is poured out of the kettle while still hot. The equation for this reaction is shown in Fig. 2-18.

The cooled solid *A* stage resin is ground and blended with enough hexamethylenetetramine to provide sufficient formaldehyde to produce a crosslinked resin. It is customary to add pigments, fillers, lime, and mold lubricants along with the hexamethylenetetramine and to heat this mixture

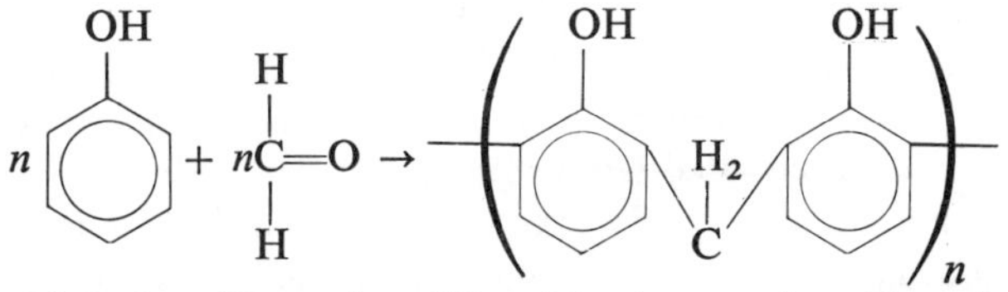

Phenol   Formaldehyde   Novolak *A* stage phenolic resin

Fig. 2-18 Formation of an *A* stage phenolic resin.

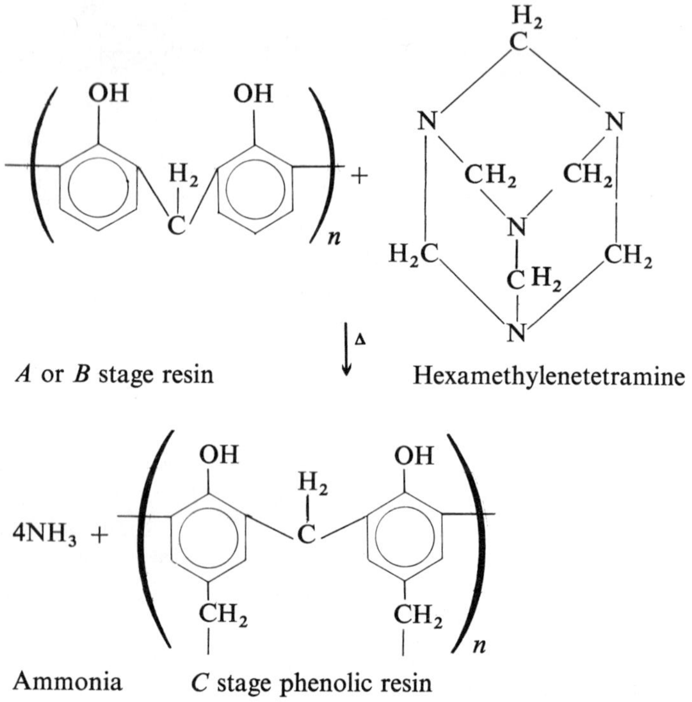

Fig. 2-19 Curing of a phenolic resin with hexamethylenetetramine.

on a two roll mill or in an extruder in order to convert the resin to a so-called B stage. The compounded mixture, which is also compacted in the advancing process, is then ground to produce a phenolic molding compound. As illustrated by the equation in Fig. 2-19, the B stage resin in the molding compound is converted to an insoluble, infusible, C stage resin in the molding process.

Urea (UF) and melamine resins (MF), produced by the condensation of formaldehyde with urea or melamine are crosslinked in the presence of phosphoric acid. Epoxy resins (EP), obtained by the condensation of bisphemol A and epichlorohydrin, are cured by polyamines such as diethylenetriamine at room temperature or dicarboxylic acid anhydrides such as phthalic anhydride at elevated temperatures. Equations for typical reactions for the formation and curing of epoxy resins (EP) are shown in Fig. 2-20.

Since unsaturated polyester resins contain ethyleneic groups such as vinyl monomers, they will copolymerize with a vinyl monomer such as styrene in the presence of an initiator such as benzoyl peroxide. Accelerators, such as N,N-dimethylaniline or cobalt naphthenate, are usually added to the

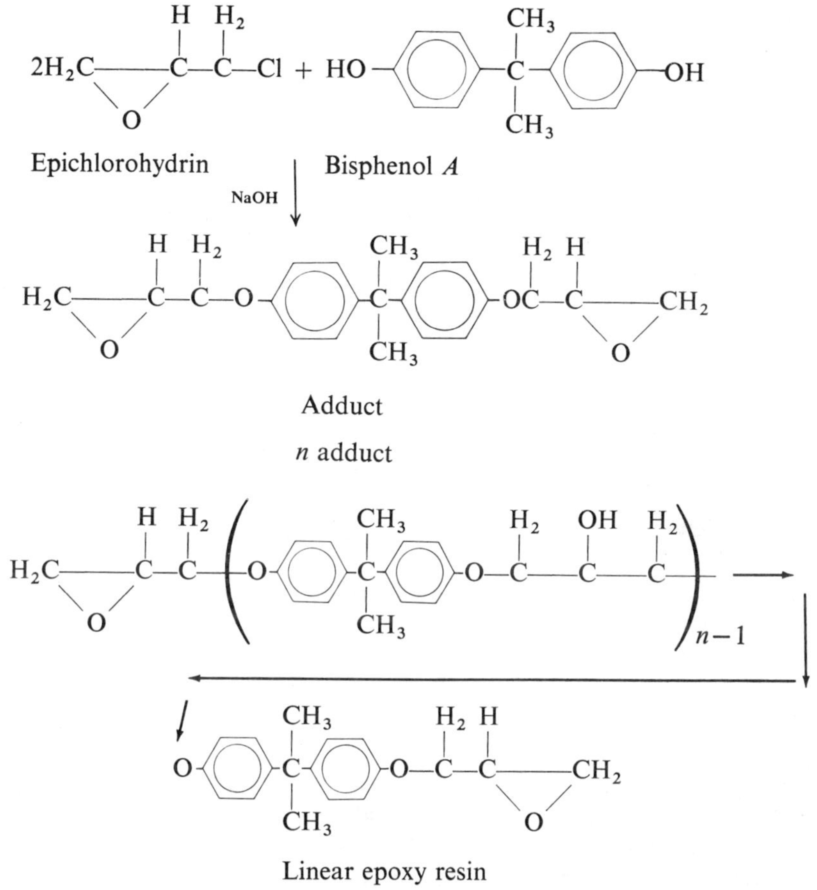

Sec. 2.4 REACTIVE ADDITIVES (CURING AGENTS)

$$2H_2C-\underset{O}{\underbrace{C}}-\overset{H}{\underset{}{C}}-\overset{H_2}{\underset{}{C}}-Cl + HO-\bigcirc-\underset{CH_3}{\overset{CH_3}{C}}-\bigcirc-OH$$

Epichlorohydrin   Bisphenol A

$\downarrow$ NaOH

Adduct

n adduct

Linear epoxy resin

Linear epoxy resin + $H_2N(CH_2)_2\overset{H}{N}(CH_2)_2NH_2 \rightarrow$ cured epoxy resin

Fig. 2-20 Equation for the formation and curing of epoxy resins.

mixture of unsaturated polyester, styrene, and benzoyl peroxide to hasten the production of phenyl radicals. Similar techniques are used for crosslinking linear allylic resins. The latter consist of a mixture of diallyl phthalate and a partially polymerized diallyl phthalate resin. The function of the initiator is exactly the same as that discussed in Sec. 2.2.

Initiators such as benzoyl peroxide may also be added to thermoplastics such as polyethylene in order to crosslink these linear polymers. Polyethylene may also be crosslinked by irradiation with gamma rays. Unsaturated polyesters may be crosslinked by irradiation with ultraviolet radiation.

2.5 COLORANTS AND PIGMENTS

Finely divided plastics scatter much of the incident light so that they are opaque. However, the extent of incident light scattering by most molded or cast plastics is sufficiently small so that these products are considered to be transparent or translucent. The transparency of poly(methyl methacrylate) (PMMA), cellulose acetate (CA), and polycarbonates (PC) are comparable to glass. These pigment-free products are often used as transparent sheets. Shellac, bitumins, and phenolic resins (PF) have characteristic colors and are also used without colorants. However, paint and plastics are usually colored or pigmented by the addition of selected additives. The presence of colorants and pigments in plants and animals is an inherent characteristic.

Paint is a mixture of a resinous vehicle and a pigment. The making of paint is an ancient art, much of which has been adopted by the plastics technologists in order to produce a wide variety of colored plastics.

The color that one sees depends on the light source or illuminant, the object, and the observer. This color may be defined in terms of the energy emitted at each wavelength for the illuminant, as the reflectance or transmittance of this energy by a plastic object, or as the effect of the intensity of the stimulus response on the observer.

Normal daylight as used in standard viewing booths, tungsten lamps, mean noon sunlight, and average daylight are used as illuminants. The last two illuminants are produced by the transmission of light from a tungsten lamp through transmission cells. The absorption and scattering of light by pigments is complex but these effects may be simplified by assuming that back scattering is predominant and that the incident light is diffuse.

Since the human eye is incapable of separating individual wavelengths of light, one sees a spectral distribution of wavelengths of light or a mixture of colors corresponding to the sum of varying amounts of the three primary stimuli, i.e., red, green, and violet. The relative importance of absorption and scattering coefficients vary with the type of colorant used. In general, organic colorants have much lower scattering coefficients than inorganic pigments.

Organic colorants are usually lower in specific gravity and provide better covering power and transparency than inorganic pigments. They may be classified in accordance with their chemical structure as azo, phthalocyanine, and anthroquinone pigments. Soluble organic colorants are called *dyes*. The term pigment is usually used to describe insoluble colorants. These colorants or toners are usually extended with finely divided colorless fillers such as calcium carbonate to produce extended colors or lakes.

Aqueous dispersions of pigments are produced by the addition of surfactants. The term *flushed color* is used to describe an aqueous dispersion of pigment and oil or resin. These dispersions are readily used but concentrated blends of color and resin are usually preferred. Dry colorants may be blended directly with polymers before calendering, extrusion, or molding.

Metallic oxides have been used as paint pigments for centuries. The most widely used metallic oxide pigment is white titanium oxide, which is available in crystalline forms as brookite, anatase, and rutile. The latter occurs naturally as large transparent crystals that are used as a substitute for diamonds in costume jewelry. However, since finely divided particles are more effective light scatterers, rutile particles in the 0.2–0.35 micron (μ) range are preferred for use as pigments. This white pigment, which is approved by the FDA, is produced at an annual rate in excess of 60 million pounds in the USA.

Ranging in color from red to yellow, naturally occurring iron oxides, called *ochres*, have been used as paint pigments for centuries. The synthetic ochres, which have a much higher tinting strength, are produced at an annual rate of 6 million pounds. Cadmium pigments, ranging in color from yellow to maroon, and lead chromate yellow are produced at annual rates in excess of 5 million pounds. However, these pigments do not meet FDA requirements. The only other widely used pigment is molybdate orange which is produced at an annual rate of 3 million pounds.

Ultramarine blue, which resembles the natural semiprecious lapis lazuli, is approved by the FDA. Other inorganic pigments are white lead molybdate, chrome oxide green, blue cobalt, blue-green hydrated chrome oxide and yellow nickel titanate. It is of interest to note that different colors are produced by coprecipitation. For example, a light-scarlet pigment is obtained by the coprecipitation of white lead molybdate, yellow lead chromate, and white lead sulfate.

2.6 FILLERS AND REINFORCEMENTS

Pigments are actually ornamental fillers and their use in paints is an old art. It is of interest to note that the pioneer plastics, celluloid and Bakelite, were filled plastics or composites. Fillers vary in size from finely divided products such as pigments, which are essentially spherical, to larger, strongly reinforcing filamentous fillers. Regardless of size, the filler in a plastic composite is the discontinuous or disperse phase present in a continuous resin matrix. There is little interaction between the discontinuous phase and the resin matrix when the former is spherical, such as in the case of glass spheres. Thus, nonreactive spherical fillers are called *extenders*.

In spite of the lack of attraction between a resin and a filler such as glass beads, the latter, available in sizes ranging from 20 to 325 mesh, may be used as a model for fillers. Actually, the maximum filler loading may be obtained by using a mixture of particle sizes consisting of 40% each of 20 and 325 mesh and 10% each of 35 and 100 mesh glass beads.

By suspending these glass spheres in viscous liquids, it can be shown that the increase in viscosity resulting from the presence of these spheres is independent of their size or the type of liquid. Actually the viscosity is dependent primarily on the concentration of the glass spheres in the liquid. This effect may be predicted by the Einstein equation which relates the viscosity of the filled (η) and unfilled liquid (η_0) to the fractional volume occupied by the filler (C), as shown in Fig. 2-21, in which the constants α and γ are 2.5 and 14.1 respectively.

$$\frac{\eta}{\eta_0} = 1 + \alpha C + \gamma C^2$$

Fig. 2-21 Modified Einstein viscosity equation.

Since physical properties such as modulus are related to viscosity (η), one may substitute many of these properties for viscosity when solid or hollow glass spheres are used as fillers in plastic composites. However, when the surface of the spheres is treated with a silane, there is an apparent reaction between the resin and the filler surface. This surface effect may be estimated by introducing a hydrodynamic factor (β) in a modified Einstein equation. In the Mooney viscosity equation, shown in Fig. 2-22, the hydrodynamic factor is 1.35 and 1.91 for closely packed and loosely packed spheres respectively.

$$\frac{\eta}{\eta_0} = \frac{\alpha C}{1 - \beta C}$$

Fig. 2-22 Mooney viscosity equation.

Many other empirical equations have been devised to correlate with experimental data. While these are beyond the scope of this discussion, it is important to note that in all equations, the mechanical properties of a composite are related to the fractional volume occupied by the filler (C) and the activity of the filler surface per unit area (SA). Thus when the filler particles are spherical and have uniform diameter, the surface activity is directly, proportional to the density (ρ) and inversely proportional to the diameter (D), i.e., $SA = 6\rho D^{-1}$.

The adsorption of a polymer on the surface of a filler is enhanced in poor solvents and the segments of the longer polymer chains are preferentially adsorbed. This absorption reduces the mobility of the polymer chain so that other segments in the chain are readily adsorbed. The optimum amount of filler is that which provides sufficient surface area for the adsorption of a monomolecular layer of polymer.

Any stress on the resin in a composite is transferred to the filler. If stress is excessive, progressive failure occurs, and the filler may be forced out of the resin matrix. When the interaction of the polymer and filler is uniform throughout the composite, the properties of the composite are said to be *isotropic*, i.e., they are the same in all directions. When the properties differ throughout the composite, the composite is said to be *anisotropic*.

Sec. 2.6   FILLERS AND REINFORCEMENTS

Soft resin matrices aid the transfer of stress to the filler. However, the application of stress usually causes flaws or cracks to propagate. Fracture toughness, increased by the substitution of filamentous fillers, reduces crack propagation.

Glass spheres are used as model fillers because their interaction with the resin matrix may be readily studied. However, most fillers, like the classical straw, hair, and cotton-fiber fillers, are filamentous. The stresses on filament-filled composites are concentrated on the ends of these fibers. The optimum reinforcing fiber has a length to diameter ratio of at least 150 to 1.

The tensile (σ_t) and flexural strengths (σ_f) of thermoplastic composites are related to the critical length of the filament (α) and its stress at failure (m'), as shown in the Kelly-Tyson equation in Fig. 2-23.

$$\sigma_t = \sigma_f C \left(1 - \frac{2}{\alpha}\right) + \alpha m'(1 - C)$$

Fig. 2-23   Kelly-Tyson equation.

For convenience, the discontinuous phase of composites is classified as a filler or fiber reinforcement. The former are listed in Table 2-2.

Many of the silica fillers are naturally occurring materials such as sand, quartz, tripoli and diatomaceous earth which differ in particle size, degree of crystallinity and hardness. Graded silica sands such as those used in hydraulic cements are also used in epoxy resin cements and in the shell-molding process. Quartz-phenolic resin composites are used as ablative insulators for nose cones, space capsules, and rocket motors.

Diatomaceous earth, also called *infusorial earth* and *fossil flour*, consists of skeletons of small organisms called *diatoms*. In contrast to sand, which has a Mohs hardness of 7, diatomaceous earth has a hardness value of less than 2 on the Mohs scale.

Synthetic silica fillers are produced by both wet and pyrolytic processes. The product from the latter processes, also called *fumed silica*, consists of loosely agglomerated particles with an average cross-section of less than 12 microns (μ). Finely divided silica fillers exhibit a graping effect that is enhanced in the presence of aliphatic polyhydric alcohols such as glycerol. These fillers are used as thixotropic agents in sheet-molding compounds (SMC) and as fillers for polyurethane foams. The reinforcing effect of these fillers is improved when the surface is treated with a silane such as vinyl tris-2-methoxysilane.

The naturally occurring silicate kaolin has been used as a filler to reduce the creep and to increase the hardness and strength of plastics. Clay, with a hardness value of 2.5 on the Mohs scale, has been used to reduce mold shrinkage and to improve the electrical properties of SMC.

Mica, nepheline syenite, talc and wollastonite are naturally occurring silicates that have been added to many plastics in order to produce composites

TABLE 2-2 Types of Fillers for Polymers

A. Silica products
  1. Minerals
      a. Sand
      b. Quartz
      c. Novaculite
      d. Tripoli
      e. Diatomaceous earth
  2. Synthetic amorphous silica
      a. Wet process silica
      b. Fumed colloidal silica
      c. Silica aerogel
B. Silicates
  1. Minerals
      a. Kaolin (China clay)
      b. Mica
      c. Nepheline syenite
      d. Talc
      e. Wollastonite
      f. Asbestos
  2. Synthetic products
      a. Calcium silicate
      b. Aluminum silicate
C. Glass
  1. Glass flakes
  2. Hollow glass spheres
  3. Solid glass spheres
  4. Cellular glass nodules
  5. Glass granules
D. Calcium carbonate
  1. Chalk
  2. Limestone
  3. Precipitated calcium carbonate
E. Metallic oxides
  1. Zinc oxide
  2. Alumina
  3. Magnesia
  4. Titania
  5. Beryllium oxide
F. Other inorganic compounds
  1. Barium sulfate
  2. Silicon carbide
  3. Molybdenum disulfide
  4. Barium ferrite
G. Metal Powders
  1. Aluminum
  2. Bronze
  3. Lead
  4. Stainless steel
  5. Zinc
H. Carbon
  1. Carbon black
      a. Channel black
      b. Furnace black
  2. Ground petroleum coke
  3. Pyrolyzed products
I. Cellulosic Fillers
  1. Wood flour
  2. Shell flour
J. Comminuted Polymers

with improved electrical and thermal properties and dimensional stability. Talc, which has a fibrous structure, has been used to produce PP composites with improved heat resistance. Wollastonite, which is acicular shaped, has been used as a filler in PVC. Asbestos is a naturally occurring, fibrous, hydrated magnesium silicate which will be discussed in the subsequent discussion of fibrous reinforcements.

Glass, available as reinforcing fibers, and as flakes, spheres, nodules and granules, has a hardness value of 4.5–6.5 on the Mohs scale. Glass flakes are obtained by smashing thin glass tubes. Composites of resins and solid glass spheres have improved mechanical properties. The addition of hollow glass spheres, called *microballoons*, yields syntactic resin foams. In contrast, the density of the composite is usually increased when solid glass fillers are added to resins. The reinforcement of glass-filled composites is improved when the glass surface is treated with a silane.

Calcium carbonate is available as wet and dry ground chalk or limestone and as precipitated calcium carbonate. The versatility and dispersibility of this filler are improved by treating the surface with resins or stearic acid. More than 1 billion lbs of this filler are used annually by the plastics industry.

Zinc oxide-polypropylene composites have improved weathering resistance. Hydrated aluminium oxide (Al_2O_3), which has a hardness value of 3 on the Mohs scale, improves the flame resistance of polyester composites. Magnesium (MgO) and titanium oxide (TiO_2) increase the stiffness and hardness of composites. Berylium oxide (BeO) microspheres increase the thermal conductivity of epoxy resin composites.

Barium sulfate ($BaSO_4$), which has a hardness value of about 3 on the Mohs scale, increases the specific gravity of plastic composites. The addition of silicon carbide (SiC), molybdenum disulfide (MoS_2) and barium ferrite ($BaFeO_4$) to resins improves the abrasive resistance, antifrictional properties, and the magnetic properties of composites respectively.

The addition of aluminum or bronze powders yields conductive composites that may be plated with other metals. Polyolefins containing finely divided heavy metals such as lead are used as neutron- and gamma-ray shields. Composites of epoxy resins and finely divided metals are used for the casting of forming tools.

Carbon black, prepared by the impingement of an oxygen-deficient oil or gas flame on a cold surface, is called *channel black*. This finely divided black filler has been replaced by a larger particle size carbon black because of the environmental pollution characteristic of the channel black process. The coarser black, with a hardness value of less than 1 on the Mohs scale and a cross-section as large as 800Å, is called *furnace black*. Crystalline polymers such as polyethylene are reinforced by carbon black. The improvement in mechanical properties is more evident when the polyethylene is crosslinked by gamma-ray irradiation.

Hollow carbon spheres produced by the vacuum pyrolysis of a slurry of resin and carbon black are used for the production of syntactic foams. Ground petroleum coke has been used as a filler for composites of polytetrafluoroethylene, and for epoxy resins.

Wood flour, obtained by the attrition mill grinding of fibrous wood, is an inexpensive filler that may be admixed with phenolic resins for the production of molding compounds. Shell flour, produced by grinding shells such as peanut shells, is less fibrous than wood flour but is used as fillers with some thermosetting plastics.

Finely divided thermoplastic polymers such as polyethylene may be added to liquid polyester resins to increase the viscosity and to provide for a smooth surface. These so-called low-profile resins usually have an index of refraction similar to that of the thermosetting resin in the composite. Foamed thermoplastic microspheres are added to other thermoplastics such as poly(vinyl chloride) in order to produce plastic composites which may be readily assembled by nailing. The wear resistance of thermoplastics may be improved by the incorporation of finely divided polyfluoroethylene (PTFE).

Synthetic rubber latex may be added to aqueous dispersions of thermoplastics such as polystyrene to produce intimate heterogenous mixtures of polymers. A tough composite is obtained when the compounded latex is coagulated. Polymers with improved impact resistance are also obtained by adding a flexible polymer such as poly(ethyl acrylate) to a rigid plastic such as poly(vinyl chloride).

Composites, usually classified as reinforced plastics, are produced by adding fibrous reinforcements to thermosetting plastics or crystalline thermoplastics. The general subject of reinforced plastics will be discussed in chapter 4 but the fibrous reinforcements will be discussed in this section. The principal types of these additives are shown in Table 2-3.

TABLE 2-3 Types of Fibrous Reinforcements for Polymers

A.	Cellulose fibers	C.	Carbon fibers
	1. Alpha cellulose		
	2. Pulp preforms	D.	Asbestos fibers
	3. Cotton flock		
	4. Jute	E.	Fibrous glass
	5. Sisal		1. Filaments
	6. Rayon		3. Chopped strand
			3. Reinforcing mat
B.	Synthetic fibers		4. Glass yarn
	1. Polyamide (nylon)		5. Glass ribbon
	2. Polyester (Dacron)		
	3. Polyacrylonitrile (Dynel, Orlon)	F.	Whiskers
	4. Poly(vinyl alcohol)	G.	Metallic fibers
	5. Other fibers		

Because of its longer fiber length, alpha cellulose produces stronger composites than wood flour. Still higher impact resistance is noted for composites filled with cotton fibers and rayon. Other cellulose fibers, such as jute and sisal, have been used as reinforcements but the properties of these composites are inferior to those obtained from rayon fibers. Powdered cellophane has also been used to reinforce high-impact polystyrene.

Nylon, poly(ethylene terephthalate) (PETP) and polyacrylonitrile (PAN) fibers have been used to reinforce thermosetting plastics. High performance composites are obtained when crystalline fibers are used as reinforcing agents. These fibers serve as nucleating agents for crystallizable polymers, such as polypropylene. Poly(vinyl alcohol) (PVAL) fibers, which are compatible with polyester and polyurethane resins, yield composites with superior properties. Strong composites are also obtained when plastics are reinforced by high modulus poly(vinyl chloride) fibers.

Carbon filaments that are superior to those originally produced by Edison by the pyrolysis of cotton fibers are now obtained commercially by the pyrolysis of rayon or polyacrylonitrile filaments. The properties of typical carbon and graphite filaments are summarized in Table 2-4. Graphite fiber composites are characterized by unusually high strength and high modulus of elasticity and by their retention of these properties to a high degree when subjected to cyclic testing.

TABLE 2-4 Typical Carbon and Graphite Filament Properties at 20°C

Filament	General Purpose Carbon	Graphite	High Modulus Graphite
Tensile strength, 10^3 psi 1-in. ga	120	90.	375
Tenacity, g/denier, 1-in. ga	6.2	5.3	15.9
Elongation at break, %	2.0	1.5	0.5
Elastic recovery, %	100	100	100
Modulus of elasticity, 10^6 psi 1-in. ga	6	6	75
Stiffness, g/denier, 1-in. ga	310	350	3,160
Density, g/cc	1.53	1.32	1.86
Equivalent diameter, microns	9.5	8.9	5.8
Volume electrical resistivity, ohm-cm $\times 10^6$	6,000	3,500	—
Surface area, m^2/g	130	4	1
Thermal conductivity, Btu/ft/(hr)/(sq ft)/(°F)	13	22	—
Specific heat, 70°F, Btu/(lb)/(°F)	0.17	0.17	0.17

As stated previously, asbestos is a naturally occurring fibrous silicate which has been used for more than 25 hundred years. There are six principal asbestiform minerals, but chrysotile is the type that is most widely used as a reinforcing agent for plastics. Approximately 2 billion lbs of chrysotile have been used annually in the USA for the production of plastic composites. Asbestos-phenolic resin composites, which have been used as rocket and missile components, are characterized by rigidity, good dimensional stability, and excellent resistance to creep at elevated temperatures.

While solid glass has been known for over 4,500 years, fibrous glass had been unknown until it was exhibited at the Paris Academy of Science about 250 years ago. The commercial process for making this filamentous product was developed in the 1930s and placed in production by the Owens-Corning Fiberglas Corp just prior to the beginning of World War II. Glass fibers may be produced from type E or C lime-aluminaborosilicate glass. Because of its lower lime content, the latter is more acid resistant then type E glass.

Type S glass has a tensile strength of 700 thousand psi while types E and C glass have tensile strengths of approximately 500 thousand psi. Type S glass also retains its physical properties at higher temperatures than ordinary glass fibers. All types of fibrous glass have large surface areas and thus are more susceptible than solid glass to deterioration by water and corrosives. The composition of the various types of fibrous glass are shown in Table 2-5.

TABLE 2-5   Composition of Types E, C and S Glass

	SiO_2	Oxides of Al and Fe	Oxides of Ca and Mg	Oxides of Na and K	B_2O_3	BaO
Type E glass	53.2	14.8	21.4	1.3	9.0	0.3
Type C glass	64.6	4.1	16.7	9.6	4.7	0.9
Type S glass	65.0	25.0	10.0	—	—	—

Glass filaments are produced by mechanically pulling molten glass after it has passed through small orifaces of controlled size. The diameter of the finished filament is controlled by the rate of winding. Multiple filaments or bundles of filaments, called *strands*, are sized and wound on forming tubes. Ropelike bundles of continuous, untwisted strands, called *rovings*, may be impregnated with a liquid resin and wound on a rotating mandrel to produce filament-wound structures. The name *pultrusion* is used to describe the process in which these impregnated rovings are forced through a die before being cured by heat.

Fibrous glass filaments are usually coated with a silane or chrome complex coupling agent to assure the formation of a good bond between the fiber and resin, as shown in the equation in Fig. 2-24. Macroradicals produced during the mechanical processing of thermoplastics also react with silane coupling agents.

The mechanical properties of thermoplastics may be improved by as much as 100% by the addition of short glass fibers. As shown by the data in Table 2-6, the coefficient of expansion of nylon is decreased and its mechanical properties are improved by the addition of fibrous glass.

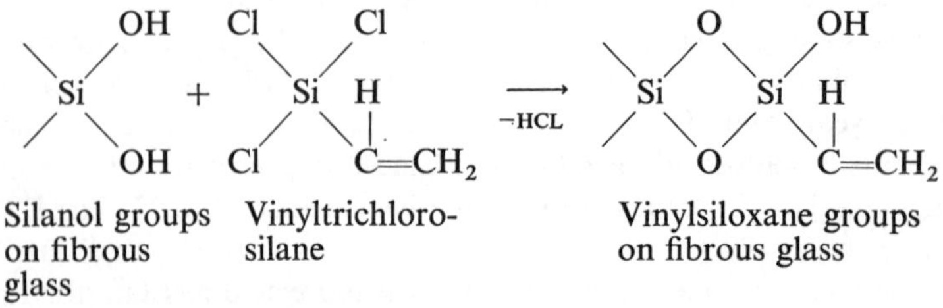

Silanol groups on fibrous glass + Vinyltrichlorosilane $\xrightarrow{-HCL}$ Vinylsiloxane groups on fibrous glass

Fig. 2-24   Equation for the reaction of a silane with surface silanol groups.

Fibrous glass mats are often used for the reinforcement of thermosetting plastics such as unsaturated polyester resins. These reinforcing mats may be produced by swirling continuous fibrous glass filaments or by bonding chopped fibers. The fibers may be bonded together by a small amount of resin or they may be held together by mechanical stitching or needling.

Sheet molding compounds may be produced by the impregnation of a fibrous glass mat with a mixture of unsaturated polyester resin and an initiator. These preimpregnated mats are cut to proper size and stacked to provide the required number of layers before heat curing in matched metal molds. This type of reinforced plastic is used for the molding of automotive bodies and components and large structural parts.

Glass yarns, obtained by twisting fibrous glass strands, are used as unidirectional reinforcements. Chopped glass strands are utilized to produce premix molding compounds called *bulk molding compounds* (BMC) and for reinforcing thermoplastics. Some structures, such as boat hulls, have been produced by the simultaneous spraying of catalyzed resin and chopped glass fibers. Unusually strong reinforcing glass fabrics have been produced by drawing parallel strands and adding a minimum number of cross-overs of the warp and woof. The more conventional glass fabrics consist of plain, taffeta, twill, and satin weaves.

Large rocket motor cases and various pressure vessels are produced by winding a preimpregnated continuous glass filament on a mandrel before curing under controlled tension. Each reinforcing strand bears an equal share of the load in these strong filament-wound composite structures.

The term *whisker* is used to describe nearly perfect single high modulus crystals with diameters in the order of a few microns. The strength of these whiskers, which is related inversely to their diameters, is attributed to the reduction of defects that contribute to the low strength of materials with

TABLE 2-6   Properties of Glass-Filled Nylons

Type Nylon	% Glass	Specific Gravity	Tensile Strength (psi)	Izod Impact (ft lbs/in)	Coefficient of Expansion $10^{-5}/in/in/°C$	Flexural Modulus (psi)	Deflection Temperature (°F @ 264 psi)
nylon-66	—	1.14	10,500	1.5	8.0	175,000	185
nylon 66	30	1.36	23,500	1.9	1.7	1,100,000	480
nylon-6	—	1.13	10,000	1.9	8.3	111,000	145
nylon-6	6	1.18	13,000	1.9	3.3	195,000	350
nylon-610	—	1.09	8,500	1.2	9.0	160,000	—
nylon-610	30	1.35	24,000	3.6	2.2	700,000	410
nylon-11	—	1.04	7,850	1.8	15.0	no break	130
nylon-11	30	1.26	12,800	2.4	3.2	no break	343
nylon-12	—	1.01	7,500	3.2	10.4	170,000	125
nylon-12	30	1.23	17,400	3.0	7.5	1,000,000	275

large cross-sections. The major defect in these small crystals is an axial screw dislocation that has slight deleterious effect on the fiber strength.

Many of these microcrystals are ceramic filaments that have been produced by vapor deposition, melt drawing, fiber spinning and extrusion. Whiskers with lengths of at least 1 ft have been produced experimentally. Maximum strength in composites is attained by uniaxial alignment relative to the applied load. However, the composites are much weaker than the reinforcing whiskers.

The principal types of whiskers are alpha aluminum oxide (Al_2O_3) or sapphire, alpha and beta silicon carbide (SiC), and silicon nitride (Si_3N_4). These extremely hard reinforcements have hardness values greater than 9 on the Moh scale. The bonding of resins to carbon, quartz, tungsten and silicon carbide filaments have been improved by growing whiskers directly on these filaments. Potassium titanate microcrystals are used as reinforcements for nylon and polyester resins.

A commercial boron filament has been produced by the hydrogen reduction of boron trichloride or by the pyrolysis of diborane on a continuous length of a 0.5 mil diameter hot tungsten wire. Carbon filaments with a diameter of 1.6 mil may also be used in place of the tungsten. A boron reinforced polyimide composite is much lighter and stiffer than aluminum.

Continuous austenitic stainless steel filaments have been used to reinforce polymers such as nylon-66 and polytetrafluoroethylene. Other metal-resin composites have been produced by the reinforcement of resins with metal fibers with a rectangular cross-section called filaments and with wire sheet.

Other additives such as stabilizers and plasticizers are discussed in Chap. 3.

BEHAVIORAL OBJECTIVES

In this chapter the usefulness of additives as compounding ingredients is discussed. After reading this chapter you should understand the following:

1. Most commercial synthetic polymers are produced by a chain reaction polymerization in which a monomer molecule adds to an initiator and the process is repeated as new monomer molecules add to the products in propagation steps.

2. The growing macromolecular chains may be terminated by coupling or by the abstraction of an atom from another molecule. The latter process is called chain transfer.

3. Chain reaction polymerization may be initiated by a free radical, an anion, a cation or by the Ziegler-Natta catalyst.

4. The half life of an initiator is the time required for one half of this reactant to be dissociated and consumed.

5. It is customary to add a small amount of an inhibitor to prevent the uncontrolled polymerization of monomers. These inhibitors are chain transfer agents.

6. Some commercial polymers such as nylon-66 are produced by a step reaction polymerization process in which all chains grow stepwise and simultaneously. Since many of these polymerization reactions are condensations, this process is sometimes called condensation polymerization.

7. It is customary to add a monofunctional compound to control the molecular weight of the polymer produced by the step reaction polymerization of bifunctional reactants. The amount of this monofunctional reactant can be calculated from a modified Carothers equation.

8. Novolac type phenolic resins are produced by the condensation of phenol with insufficient formaldehyde in the presence of sulfuric acid. The linear or A-stage resin is advanced to a B-stage and finally crosslinked in the presence of hexamethylenetetramine which supplies more formaldehyde.

9. Thermoplastic urea and melamine resins are crosslinked by heating with acids.

10. Epoxy resins are cured at ordinary temperatures by the addition of polyamines. Anhydrides of dicarboxylic acids are used for curing EP at higher temperatures.

11. Solutions of unsaturated polyesters in styrene are cured by initiators such as benzoyl peroxide. The rate of crosslinking may be accelerated by the use of N-N-dimethylaniline or cobalt naphthenate.

12. Inorganic pigments or organic colorants are admixed with resins to produce plastics with specified colors.

13. A filler which is called the discontinuous phase is added to a resin which is called the binder, matrix or continuous phase to increase the hardness and modulus of the composite or to extend the resin.

14. The stress applied to a resin in a composite is transferred to the filler and this is effected best by soft resin matrices.

15. Many naturally occurring and synthetic silicas and silicates are used as fillers for resins.

16. Glass which is one of the most widely used fillers is available as flakes, spheres, nodules, granules as well as fibrous glass.

17. Other widely used fillers are limestone, powdered aluminum, carbon black and wood flour.

18. Reinforced plastics contain fibrous reinforcing agents such as alpha cellulose, synthetic fibers, fibrous glass or whiskers.

19. The extent of adhesion between the surface of fibrous glass and the resin is improved by treating the surface with a coupling agent such as a silane finish.

20. Strong composites are produced by the filament winding process in which preimpregnated glass filaments are wound on a mandrel and cured.

21. Whiskers which yield extremely strong composites are single microcrystals produced by special techniques.

GLOSSARY

A-stage resin: a thermoplastic phenolic resin.
Acicular: needlelike.
Alpha cellulose: cellulose that is insoluble in a 17.5 percent sodium hydroxide solution.
Anion: a negatively charged atom.
Anisotropic: properties directionally dependent.
C-stage resin: a thermoset phenolic resin.
CA: cellulose acetate.
Carother's equation: $\overline{DP} = 1/1 - P$ where P is equal to the extent of polymerization.
Cation: a positively charged atom.
Chain reaction polymerization: a process in which a monomer adds to a free radical or ionic initiator and this step is repeated over and over again as new monomer molecules add to the growing chain. This process is sometimes called addition polymerization.
Chain transfer: a process in which a growing polymer chain abstracts an atom from another molecule to produce a dead polymer and a new active center.
Chain transfer constant: a measure of the effectiveness of a chain transfer agent.
Diatomaceous earth: finely divided silica consisting of the skeletons of diatoms.
Einstein's equation: a correlation of the viscosity increase and the fractional volume occupied by a filler in a composite.
EP: epoxy resin.
Fibrous filler: one in which the ratio of the length to diameter is at least 150 to 1.
Inhibitor: a chain transfer agent which yields a dead polymer and an inactive free radical.
Isotropic: having similar properties in all directions.
Kaolin: clay.
Low profile resins: resins such as unsaturated polyesters containing finely divided PE filler.
MF: melamine resin.
Microballoons: hollow glass spheres.
Modulus: the ratio of the elongation to the applied stress.
Mohs hardness: a scale of hardness ranging from 10 for diamonds to 1 for talc. See F-18 "Handbook of Chemistry and Physics," The Chemical Rubber Co., Cleveland, 1975.

Mooney equation: a modified Einstein equation which includes a hydrodynamic factor to account for surface effects of fillers.
Novolac: a phenolic resin made by a two-step process in which the crosslinking of the thermoplastic resin takes place in the mold.
NR: natural rubber.
Oligomer: low molecular weight polymer.
PA: nylon.
Paint: a mixture of a resinous vehicle and a pigment.
PAN: polyacrylonitrile.
PC: polycarbonate.
Peptizers: chain transfer agents used to assure degradation of polymers during mechanical processing.
PETP: poly(ethylene terephthalate).
PUR: polyurethane.
PVAC: poly(vinyl acetate).
PVAl: poly(vinyl alcohol).
Radical: an electron deficient molecule.
Resin: an uncompounded plastic.
Roving: a bundle of untwisted strands.
SMC: sheet molding compound.
Step reaction polymerization: a polymerization process in which chain growth occurs stepwise as all chains grow simultaneously.
Strand: a bundle of glass filaments.
Syntactic foam: a composite of resin and hollow glass spheres.
Telogen: a chain transfer agent.
Type C glass: acid resistant glass used for fibers.
Type E glass: electrical grade glass used for fibers.
UF: urea-formaldehyde resin.
Whiskers: single crystals used as reinforcements.
Yarn: a bundle of twisted strands.
Ziegler, Karl: a German scientist who shared the Nobel Prize with Giulio Natta for the synthesis of HDPE.
Ziegler-Natta catalyst: a mixture of a titanium chloride and an aluminum alkyl.

SUGGESTED QUESTIONS

1. What are the three steps in chain polymerization?
2. What is meant by the term chain transfer?
3. Why is it incorrect to call initiators such as benzoyl peroxide and potassium persulfate catalysts?
4. What prevents a vinyl monomer such as styrene from polymerizing spontaneously?

COMPOUNDING INGREDIENTS Chap. 2

5. Why is acetic acid added to the bifunctional reactants (adipic acid and hexamethylenediamine) in the production of nylon-66?
6. What curing agent is used to convert an A stage novolac resin to the C stage or crosslinked resin?
7. What curing agents are used to cure EP?
8. What curing agents are used to cure unsaturated polyester?
9. Why is a pigmented plastic superior to a painted article of similar shape?
10. Which is preferred for pastel colors, PF or PU?
11. Why are glass spheres preferred as an extender-type filler?
12. Which shape of filler will improve the physical properties of a composite more, a spherical or a nonspherical filler?
13. What type of filler will produce the tougher composite, a spherical or a filamentous filler?
14. Name three silica fillers.
15. Name three silicate fillers.
16. What filler would you use to obtain a low density composite?
17. Why is wood flour preferred over other commercial cellulosic fillers?
18. What filler would you use to obtain a smooth surface in a reinforced polyester?
19. Which additive will produce the stronger composite, carbon black or graphite fibers?
20. What is the principal difference between type E and type C glass?
21. Which is stronger, a composite made by curing SMC or a filament wound structure?
22. What is the composition of a boron reinforcing filament?

ANSWERS

1. Initiation, propagation and termination.
2. The process in which a growing polymer chain abstracts an atom from another molecule to produce a dead polymer and a new active center.
3. Because a fragment of these initiators is present on the end of the polymer chain and remains in the polymer.
4. An inhibitor which acts as a chain transfer agent.
5. To control the molecular weight.
6. Hexamethylenetetramine which yields formaldehyde when heated.
7. Polyamines are used at ordinary temperatures and anhydrides of dicarboxylic acids are used at higher temperatures.
8. Free radical initators such as benzoyl peroxide.
9. The color in plastics is distributed uniformly throughout the plastic article.
10. PU since PF is dark colored.
11. Large volumes can be used because of little interaction between this filler and the resin matrix.

12. Nonspherical cause a greater reduction of the polymer mobility.
13. Filamentous.
14. Quartz, novaculite, fumed silica, etc.
15. Clay, mica, aluminum silicate, etc.
16. A low density filler such as hollow glass spheres or microballoons.
17. Wood flour is more fibrous than other products such as ground shell fillers.
18. Finely divided PE.
19. Graphite fibers.
20. Their chemical composition. Type C is more resistant to acids.
21. The filament wound composite.
22. It consists of a shell of boron deposited on a tungsten or carbon filament core.

chapter 3

Stabilizers, Plasticizers, and Other Additives

3.1 ANTIOXIDANTS

Polymers formed by the chain reaction polymerization of vinyl monomers such as polypropylene (PP) are particularly susceptible to oxidative degradation. Polymers formed by step reaction polymerization such as nylon-66 (PA) are also subject to this type degradation but to a lesser extent than those produced by chain reaction or addition polymerization.

It is generally accepted that the carbon-hydrogen bonds in polymer chains may be cleaved in the presence or absence of oxygen by the mechanisms shown by the equations in Fig. 3-1. Because of the greater stability of a tertiary free radical, that is, one that lacks hydrogen atoms on the electron deficient carbon atom, polymers such as polypropylene, which contain tertiary carbon atoms, readily undergo the initiation step illustrated in Fig. 3-1. This step, which is catalyzed by heavy metals, occurs more readily in the amorphous or less ordered regions of a polymer system.

Sec. 3.1 ANTIOXIDANTS 65

$$P:H \xrightarrow{\Delta} H\cdot + P\cdot$$
Polymer   Hydrogen   Macroradical

or

$$P:H + O:O \rightarrow P:O:OH \rightarrow \cdot OH + P:O\cdot$$
Polymer Oxygen   Peroxide   Hydroxyl   Oxymacro-
                                       radical   radical

Fig. 3-1 Initiation step in polymer degradation.

$$P:P \xrightarrow{\Delta \text{ or } E} P\cdot + P\cdot$$
Polymer   Macro-   Macro-
                  radical radical

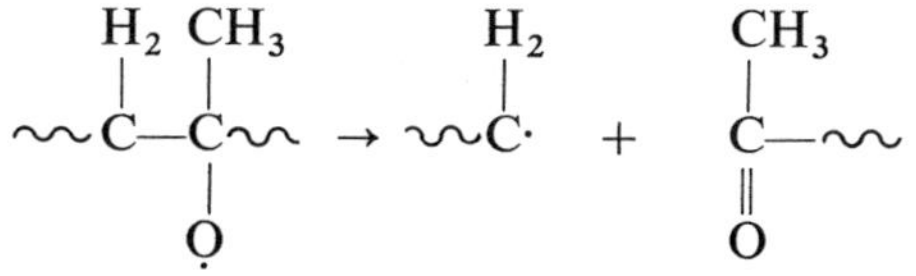

$$\begin{array}{c} H_2\ CH_3 \\ |\ \ \ | \\ \sim\!\!C\!-\!\!C\!\sim \\ | \\ O \end{array} \rightarrow \begin{array}{c} H_2 \\ | \\ \sim\!\!C\cdot \\ \end{array} + \begin{array}{c} CH_3 \\ | \\ C\!-\!\!\sim \\ \| \\ O \end{array}$$

Oxymacro-   Macro-   Polymer
radical   radical

Fig. 3-2 Macroradical formation by homolytic cleavage.

Homolytic cleavage resulting from mechanical, ultraviolet, or thermal degradation may also yield macroradicals. Figure 3-2 shows that this type of chain scission also takes place with both polymers and oxymacroradicals.

Figure 3-3 shows that free radicals produced either by chain scission or in the initiation step may propagate by reacting with oxygen or by abstracting a hydrogen atom from a neighboring polymer chain.

$$P\cdot + O_2 \rightarrow POO\cdot$$
Macro-   Oxygen   Peroxy-
radical   macro-
                  radical

$$POO\cdot + PH \rightarrow POOH + P\cdot$$
Peroxy-   Polymer   Peroxide   Macro-
macro-   radical
radical

$$PO\cdot + PH \rightarrow POH + P\cdot$$
Oxymacro- Polymer   Polymer   Macroradical   radical

$$\cdot OH + PH \rightarrow HOH + P\cdot$$
Hydroxyl   Polymer   Water   Macroradical   radical

Fig. 3-3 Propagation steps in polymer degradation.

When these macroradicals are mobile, they may couple to form stable polymers. Coupling may produce network polymers when more than one unpaired electron is present on the same polymer chain. Typical equations for termination by coupling are shown in Fig. 3-4.

Propagation and coupling of radicals may be prevented if the radical abstracts a hydrogen atom from an additive, called an *antioxidant*, in a process called *chain transfer*. Many of the commercial antioxidants (AH) such as 2,6-ditert. butyl-4-methylphenol and 2,2'-methylene-bis-(4-methyl-6-tert. butylphenol) are hindered phenols. As shown in Fig. 3-5, the new free radicals formed by chain transfer are stabilized by resonance and are incapable of propagation or reaction with oxygen.

Additional antioxidants, called *synergists* or *peroxide decomposers*, are often used in conjunction with the hindered phenols, called *primary antioxidants*. Synergists may be sulfides (RSR), such as dilauryl thiodipropionate, or phosphites $((RO)_3P)$, such as tris-nonylphenyl phosphite. Figure 3-6 shows that these synergists may abstract oxygen from peroxymacroradicals.

$$P\cdot \;+\; \cdot P \;\rightarrow\; P:P$$

Macro-   Macro-   Polymer
radical   radical

$$\sim\sim C\cdot \;+\; \cdot C\sim\sim \;\rightarrow\; -C:C\sim\sim$$
(with H_2 on each carbon)

Macro-   Macro-   Polymer
radical   radical

$$PO\cdot \;+\; \cdot OP \;\rightarrow\; POOP$$

Oxymacro-   Oxymacro-   Polymer
radical   radical

$$P\cdot \;+\; \cdot OH \;\rightarrow\; POH$$

Macro-   Hydroxyl   Polymer
radical   radical

$$POO\cdot \;+\; POO\cdot \;\rightarrow\; POOOOP \;\rightarrow\; POOP \;+\; O_2$$

Peroxy-   Peroxy-   Polymer   Polymer   Oxygen
macro-   macro-
radical   radical

$$POO\cdot \;+\; PO\cdot \;\rightarrow\; POOOP \;\rightarrow\; POP \;+\; O_2$$

Peroxy-   Oxymacro-   Polymer   Polymer   Oxygen
macro-   radical
radical

Fig. 3-4 Termination of radicals by coupling.

ANTIOXIDANTS

$$P\cdot + AH \rightarrow PH + A\cdot$$

Macro-radical — Antioxidant — Polymer — Stable radical

$$POO\cdot + AH \rightarrow POOH + A\cdot$$

Peroxy-macroradical — Antioxidant — Polymer — Stable radical

$$PO\cdot + AH \rightarrow POH + A\cdot$$

Oxymacroradical — Antioxidant — Polymer — Stable radical

$$HO\cdot + AH \rightarrow HOH + A\cdot$$

Hydroxyl radical — Antioxidant — Water — Stable radical

Fig. 3-5 Chain transfer of macroradicals with antioxidants.

Figure 3-7 shows that hindered phenols may be regenerated in the presence of these synergists.

Polymers such as polyethylene are somewhat resistant to degradative oxidation and hence require only traces of antioxidants to prevent oxidation during melt processing and when used in outdoor service. Carbon black is often added to polyethylene as a pigment and stabilizer. However, this additive often deactivates hindered phenols and therefore other antioxidants such as thiobisphenols are generally used in the presence of carbon.

$$POO\cdot + (RO\overset{O}{\underset{\|}{C}}(CH_2)_n)_2S \rightarrow P\cdot + (RO\overset{O}{\underset{\|}{C}}(CH_2)_n)_2\underset{\underset{O}{\|}}{\overset{O}{\underset{\|}{S}}}$$

Peroxy-macroradical — Dialkylthioester — Macroradical — Sulfone

$$POO\cdot + 2(RO)_3P \rightarrow P\cdot + 2(RO)_3P{=}O$$

Peroxy-macroradical — Trialkylphosphite — Macroradical — Trialkylphosphate

Fig. 3-6 Peroxymacroradical decomposition by synergists.

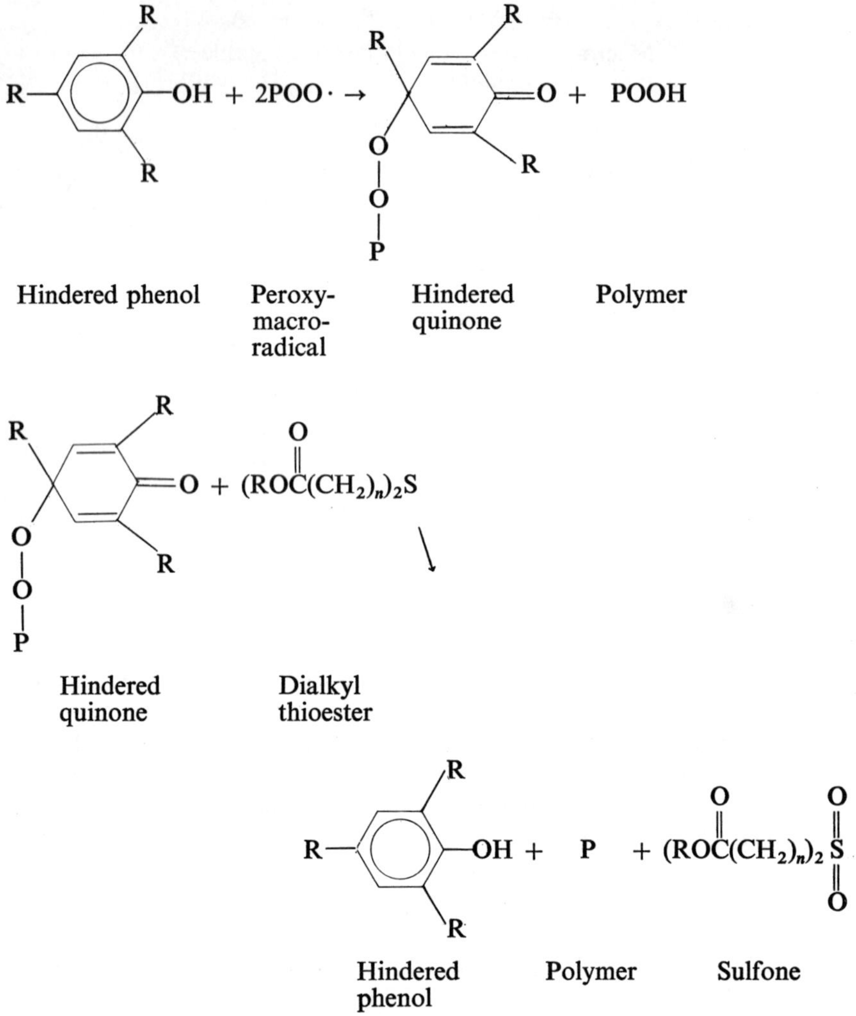

Fig. 3-7 Regeneration of hindered phenols by synergists.

Polymers, such as polypropylene, containing tertiary carbon atoms $\left(\begin{array}{c}H\\|\\-C-\\|\\R\end{array}\right)$ are more readily degraded than those containing secondary carbon atoms $\left(\begin{array}{c}H\\|\\-C-\\|\\H\end{array}\right)$, such as those present in polyethylene. Polymers, such as rubber-modified polystyrene, containing unsaturated groups $\left(\begin{array}{cc}H&H\\|&|\\-C=C-\end{array}\right)$

are particularly sensitive to degradation during thermal processing and in outdoor service and require larger amounts of antioxidants.

The use of antioxidants and synergists may be illustrated by the following formulations for polypropylene. This polymer degrades appreciably after 300 hrs at 300°F when either 0.5 parts of a hindered phenol or 0.5 parts of a dialkyl thioester are added to 100 parts of resin. However, when 0.1 PPH of a hindered phenol and 0.3 PPH of a dialkyl thioester are present together, the compounded polymer does not degrade appreciably after 1800 hrs at 300°F.

3.2 HEAT AND ULTRAVIOLET STABILIZERS

Chlorine-containing polymers such as poly(vinyl chloride), vinyl chloride copolymers, and poly (vinylidene chloride) tend to lose hydrogen chloride when heated or exposed to ultraviolet (UV) radiation. Hydrocarbon polymers containing tertiary carbon atoms such as polypropylene and polystyrene are also susceptible to thermal and UV degradation. The energy in the UV region (290–400 millimicrons (mμ) is 71–95 kcal per mole and this is sufficient to break carbon-carbon, carbon-chlorine, and carbon-hydrogen bonds. Therefore, it is essential that these high energy photons (hν) be either filtered out or screened by pigments or stabilizers. Carbon black, calcium carbonate, and titanium dioxide are often used as pigments in PVC.

Antioxidants, discussed in the previous section, are usually effective heat and ultraviolet stabilizers for polypropylene. However, specific UV stabilizers must also be added to PVC and other chlorine-containing plastics.

The UV decomposition of PVC is autocatalytic. That is, the first molecule of HCl that is released catalyzes additonal decomposition in what is called a *zipper* reaction. As shown by the equation in Fig. 3-8, ethylene

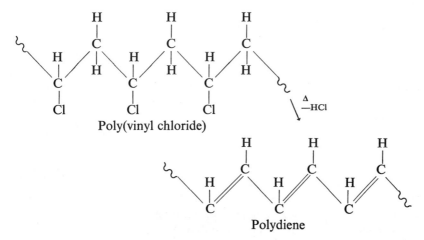

Fig. 3-8 Equation for the degradation of poly(vinyl chloride).

groups $\left(\begin{array}{cc} H & H \\ | & | \\ -C=C- \end{array}\right)$ are produced in this reaction. Since these are chromophoric or color-producing groups, the color of the degradation product changes progressively from yellow to red to brown and finally to black as the extent of unsaturation in the polydiene product increases.

Heavy metal salts or soaps, which are also used as lubricants, may react with HCl and hence are effective scavengers. However, many of these additives and end products are incompatible and exude or "plate out." Therefore, more compatible phenolic complexes of barium, cadmium, or zinc are often used. Organotin compounds, such as dibutyltin dilaurate, are often recommended. Equations for the reaction of HCl with soaps and with epoxy stabilizers are shown in Fig. 3-9.

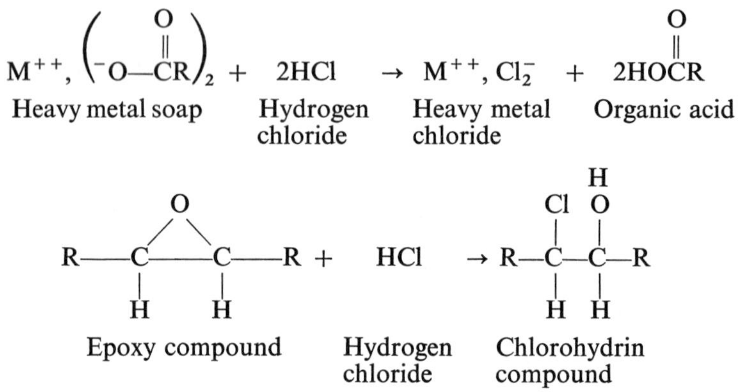

$$M^{++}, \left(\begin{array}{c} O \\ \| \\ ^-O-CR \end{array}\right)_2 + 2HCl \rightarrow M^{++}, Cl_2^- + 2HOCR\overset{O}{\|}$$

Heavy metal soap   Hydrogen chloride   Heavy metal chloride   Organic acid

$$R-\underset{H}{\overset{O}{\overset{\triangle}{C}}}-\underset{H}{C}-R + HCl \rightarrow R-\underset{H}{\overset{Cl}{C}}-\underset{H}{\overset{OH}{C}}-R$$

Epoxy compound   Hydrogen chloride   Chlorohydrin compound

Fig. 3-9   Hydrogen chloride scavenger reactions.

Organic phosphites are usually added to heavy metal soaps for the stabilization of PVC. Lead salts have been used for some industrial and electrical applications but are not acceptable for use in applications such as PVC potable water pipes. Organic phosphites and epoxy compounds are acceptable for food applications but they are not effective unless accompanied by alkyltin salts or calcium and zinc soaps. It is interesting to note that PVC has been known for over one hundred and fifty years but its commercial use was severely limited until the development of efficient organotin mercaptide stabilizers.

Nickel complexes of thiobisphenols have also been used for the outdoor stabilization of poly(diethyl p-methoxybenzylidene malonate) and 2,4,6-tritert. butylphenol is particularly effective for the stabilization of polystyrene which absorbs radiation in the 300–320 mμ range.

The classical sunscreen or UV stabilizer was phenyl salicylate. Figure 3-10 shows that this ester rearranges in the presence of UV radiation to pro-

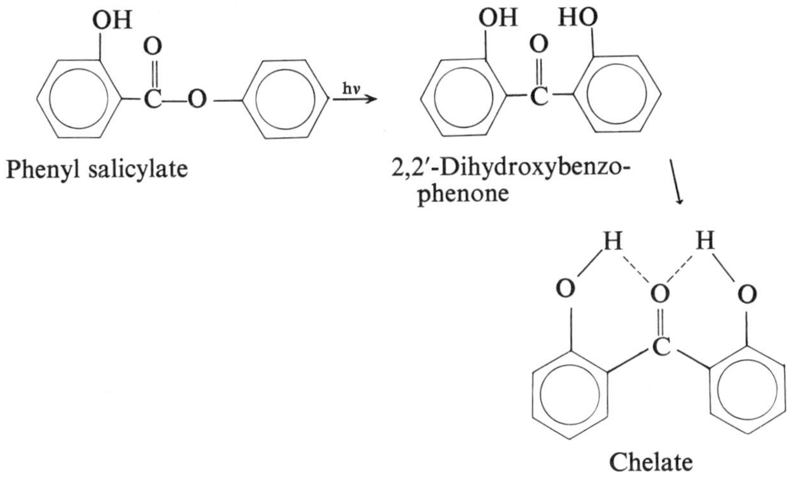

Fig. 3-10 Rearrangement of phenyl salicylate in the presence of ultraviolet radiation.

duce a 2,2'-dihydroxybenzophenone which, by hydrogen bonding, forms a cyclic structure called a chelate. It is believed that high energy photons (hv) are absorbed by these chelates and that less harmful lower energy photons are released.

The most widely used UV absorbers are 2-hydroxybenzophenones, 2-hydroxyphenylbenzotriazoles and 2-cyanodiphenyl acrylates. These stabilizers like the product produced by the rearrangement of phenyl salicylate absorb photons (hv) and form chelates in the 290–400 mμ range. The chelate formation, subsequent rearrangement of 2,4-dihydroxybenzophenone and the release of lower energy photons (hv') are shown in Fig. 3-11.

3.3 PLASTICIZERS

The two most important advances in polymer technology in the nineteenth century are Charles Goodyear's use of sulfur to cure natural rubber and John W. Hyatt's use of camphor to increase the flexibility of cellulose nitrate. The latter technique, called *plasticization*, is essential for the use of cellulose nitrate and poly(vinyl chloride) as moldable plastics.

Flexibilization of polymers of vinyl chloride may also be accomplished by the copolymerization of vinyl chloride (VCM) with monomers such as vinyl acetate or ethyl acrylate. Such copolymerization is sometimes called internal plasticization to differentiate it from the external plasticization observed when high boiling substances, such as dioctyl phthalate, are added

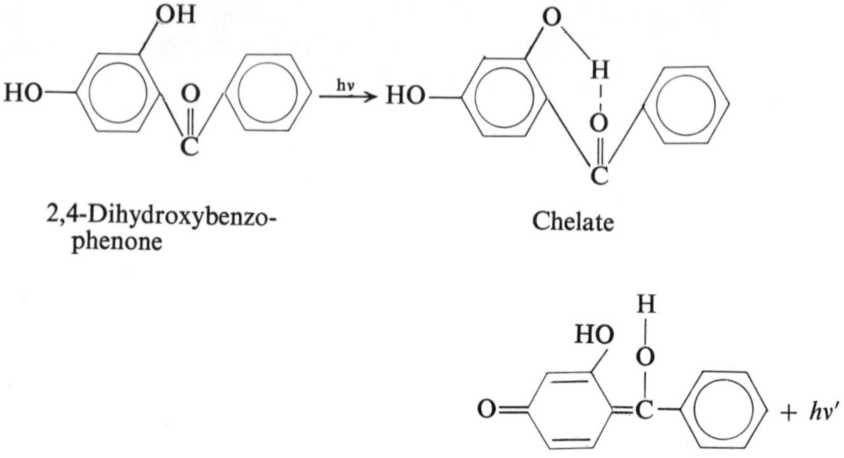

2,4-Dihydroxybenzophenone

Chelate

Rearranged quinoid form

Fig. 3-11 Proposed equation for the absorption of high energy photons (hν) and release of lower energy photons (hν′) by a hydroxybenzophenone.

to rigid plastics. An *external plasticizer* is a nonvolatile additive with a solubility parameter value similar to that of the polymer.

The use of plasticizers has grown from the few pounds of camphor used by Hyatt in 1868 to an annual world wide use of over 2.25 billion lbs. The annual production of plasticizers in the USA is in excess of 1.25 billion lbs. Esters of phthalic acid account for almost two-thirds of this volume. More than 80% of these plasticizers are used for the plasticization of poly(vinyl chloride). Dioctyl phthalate is the most widely used external plasticizer.

It is assumed that a plasticizer, like a solvent, reduces some of the intermolecular secondary valence bonds and thus facilitates the mobility of rigid polymer chains. Because of strong dipole-dipole interaction of the methylene (CH_2) and chloromethyl (CCl) groups on neighboring chains, the intermolecular forces or specific cohesive forces of poly(vinyl chloride) are greater than 2500 cal/per/mole. This polymer has a solubility parameter value of 9.7H. The values of the more widely used plasticizers, dioctyl sebacate, dioctyl phthalate, and tricresyl phosphate, are 8.7H, 8.9H and 9.8H respectively.

The addition of a small amount of plasticizer actually increases the hardness of poly(vinyl chloride) but this hardness is reduced by 5, 20 and 35% by the addition of 30, 50, and 70 PPH respectively of dioctyl phthalate. The glass transition temperature (T_g) of poly(vinyl chloride) is also reduced by the addition of relatively large amounts of plasticizers. The term *plasticizer efficiency* is used to describe the effectiveness of a plasticizer in reducing the hardness and tensile strength and in increasing the flexibility of a plastic. This parameter is related to the relative amounts of different plasticizers that must be used to achieve a similar degree of plasticity.

Esters of high molecular weight monocarboxylic acids, such as octyl stearate, contribute good low-temperature properties, but esters of noncyclic dicarboxylic acids, such as dioctyl adipate, are preferred for low temperature applications of poly(vinyl chloride). Tricresyl phosphate, which was the pioneer plasticizer for PVC, provides flexible products characterized by good flame and heat resistance. Esters such as tri-2-ethylhexyl phosphate also contribute to good low-temperature properties.

The solubility parameter value of dioctyl phthalate is similar to that of mineral oil and thus this plasticizer may be extracted by mineral oil. In contrast, polymeric plasticizers that are produced by the esterification of dicarboxylic acids and dihydric alcohols are less toxic, nonvolatile, and resistant to extractions by mineral oil. Epoxidized vegetable oils are good plasticizers as well as heat and light stabilizers.

Plasticizers usually cost considerably more than PVC. Less expensive oils, called *extenders*, are sometimes used in conjunction with the primary plasticizer to reduce costs. The solubility parameter value of the extender is usually outside the desired range, but the value for the mixture should be in the 9.4–9.8H range to insure compatibility. Structural formulas for typical plasticizers are shown in Fig. 3-12.

Thin, rigid molecules, such as chlorinated biphenyl esters of abietic acid and diphenyl phthalate, are antiplasticizers for polycarbonates, polysulfones, cellulose triacetate, and poly(vinyl chloride). Antiplasticizers are rigid molecules that actually increase the hardness and modulus and reduce the flexibility of rigid plastics.

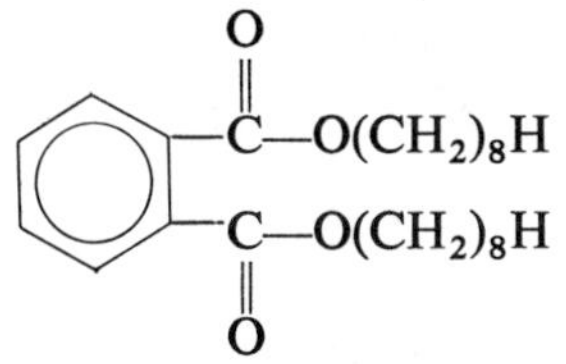

Dioctyl phthalate

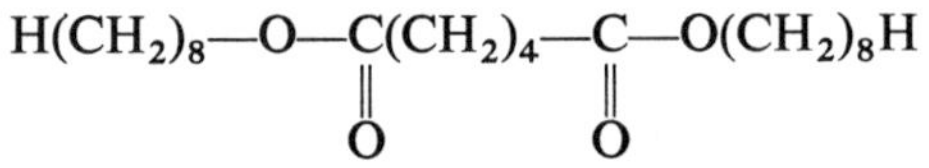

Dioctyl adipate

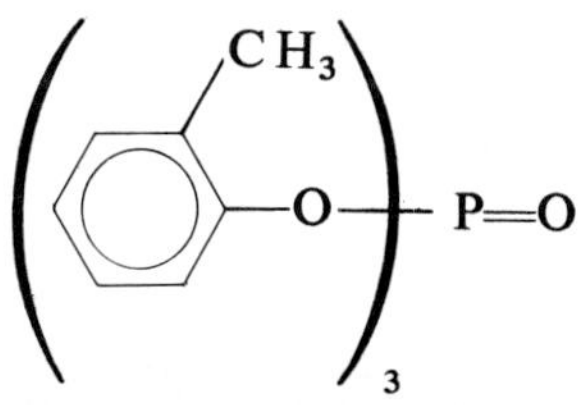

Tricresyl phosphate

Fig. 3-12 Formulas for typical plasticizers.

3.4 FLAME RETARDANTS

In spite of increased safety measures, the frequency of fires per million persons has not decreased and so the annual dollar loss due to fires has continued to increase during the past decade. The loss attributable to fires was greater than $2.5 billion in 1974. Since the use of plastics as materials of construction continues to increase, it is imperative that these economical materials not contribute to this loss. The FHA and the Hill-Burton Program require that only flame-resistant materials of construction be used in resident housing and in hospital and nursing home construction respectively. The flame spread and smoke density of PVC sheet flooring and asbestos filled tile is similar to that of wool carpeting and less than that of red oak flooring but the flame spread and smoke density of nylon and acrylic carpets are similar to that of red oak flooring.

Building materials will burn if the temperature exceeds the ignition temperature, if combustible gases are volatilized, or if oxygen is not excluded from these materials at elevated temperatures. If combustible plastics, red oak flooring, or wool carpet are allowed to burn, carbon monoxide and carbon dioxide will be produced. Burning wool and acrylic carpets and PVC plastics produce hydrogen cyanide (HCN) and hydrogen chloride (HCl), respectively.

As shown in Fig. 3-13, combustion is a chain reaction that is initiated and propagated by hydroxyl free radicals ($\cdot$OH). It is assumed that these free radicals are produced by chain transfer of oxygen with alkyl radicals ($RCH_2\cdot$) and that halogen free radicals ($X\cdot$) terminate this chain reaction. Hence, it is not surprising that some halogen compounds are flame retardants.

$$\underset{\substack{\text{Hydroxyl} \\ \text{radical}}}{\cdot OH} + \underset{\substack{\text{Hydro-} \\ \text{carbon}}}{RCH_2H} \rightarrow \underset{\substack{\text{Alkyl} \\ \text{radical}}}{RCH_2\cdot} + \underset{\text{Water}}{HOH}$$

$$\underset{\substack{\text{Alkyl} \\ \text{radical}}}{RCH_2\cdot} + \underset{\text{Oxygen}}{O\!-\!O} \rightarrow \underset{\text{Aldehyde}}{RCHO} + \underset{\substack{\text{Hydoxyl} \\ \text{radical}}}{\cdot OH}$$

$$\underset{\substack{\text{Hydrogen} \\ \text{halide}}}{HX} + \underset{\substack{\text{Hydroxyl} \\ \text{radical}}}{\cdot OH} \rightarrow \underset{\text{Water}}{HOH} + \underset{\substack{\text{Halogen} \\ \text{radical}}}{X\cdot}$$

$$\underset{\substack{\text{Halogen} \\ \text{radical}}}{X\cdot} + \underset{\substack{\text{Alkyl} \\ \text{radical}}}{RCH_2\cdot} \rightarrow \underset{\substack{\text{Alkyl} \\ \text{halide}}}{RCH_2X}$$

Fig. 3-13   Equation for combustion process in presence of halides.

Much of the present knowledge of flame retardants is based on empirical tests. Tests for flammability are also empirical. While antimony trioxide (Sb_2O_3) is not an effective flame retardant, antimony oxychloride (SbOCl) or equimolar mixtures of antimony trioxide and halogen compounds are effective flame retardants. It is assumed that the hydrogen halide (HX) produced by the thermal decomposition of the halogen compounds reacts with the antimony trioxide to produce antimony trihalide (SbX_3) or antimony oxyhalide.

These mixtures are called *synergistic flame retardants.* Mixtures of 5% antimony trioxide and 10% bromine are equivalent in flame retardancy to those containing 14% chlorine. The oxyhalides are thermally decomposed to produce antimony trioxide. Suggested equations for these thermal reactions are shown in Fig. 3-14.

It is assumed that antimony oxyhalides provide a heavy gaseous insulating barrier that prevents the reaction of oxygen with the plastic surface. A strong char structure also develops in the presence of antimony halides, causing rapid extinguishment of the flames. It is of interest to note that while a polyester casting resin containing 11.5% of tetrabromophthalic anhydride did not char when burned, charring but no burning was observed when 5% antimony oxide was added to this mixture.

Superior flame retardancy is also noted when phosphorus compounds such as tricresyl phosphate are added to halogen compounds such as chlorinated paraffin wax. It is assumed that less volatile phosphorus halides are

$$Sb_2O_3 + 6HX \rightarrow 2SbX_3 + 3H_2O$$
Antimony trioxide + Hydrogen halide → Antimony trihalide + Water

$$Sb_2O_3 + 2HX \rightarrow 2SbOX + H_2O$$
Antimony trioxide + Hydrogen halide → Antimony oxyhalide + Water

$$5SbOX \xrightarrow{490-535°F} SbX_3 + Sb_4O_5X_2$$
Antimony oxyhalide → Antimony trihalide + An antimony oxyhalide

$$4Sb_4O_5X_2 \xrightarrow{770-890°F} 5Sb_3O_4X + SbX_3$$
An antimony oxyhalide → An antimony oxyhalide + Antimony halide

$$3Sb_3O_4X \xrightarrow{890-1050°F} 4Sb_2O_3 + SbX_3$$
An antimony oxyhalide → Antimony trioxide + Antimony trihalide

Fig. 3-14 Thermal reactions of antimony oxide and halides.

produced and that these reduce combustion at the char zone. Since phosphorus tribromide (PBr_3) is less volatile than phosphorus trichloride (PCl_3), bromocompounds are more effective in this synergistic reaction. Some degree of flame retardancy is also provided by boron and nitrogen compounds.

While most flame retardants are nonreactive additives, reactive chloro or phosphorus compounds are also used. For example, tetrachlorophthalic anhydride, tetrabromobisphenol-A and phosphorus containing dihydric alcohols are used to produce flame retardant polyesters, epoxy resins, and polyurethanes, respectively.

Over 100 million lbs of flame retardants are consumed annually by the American plastics industry. Because of the continued growth of plastics as materials of construction and emphasis on safety, use will continue to grow. Both chlorinated paraffin wax and antimony oxide are available as solids and as aqueous dispersions.

3.5 MISCELLANEOUS ADDITIVES

Static electricity builds up on the surface of nonpolar plastics, such as polyethylene, during processing. In the absence of contact with conductors such as metals or antistatic agents, this charge increases at various stages of production, distribution, and end-use of plastic parts. Since this static charge attracts dust, a cleaning step may be necessary before shipment, before purchase by the consumer, and continuously during use. Classical dusting techniques may remove the surface dust temporarily, but dusting with a rag actually increases the static charge and aggravates the condition. A build up of 3,000V of static electricity in a plastic container is sufficient to ignite gasoline or combustible solvents.

Antistatic agents may be incorporated in the plastic part or they may be applied externally to the plastic surface. These agents may be amines (RNH_2), quaternary ammonium salts (R_4N^+, X^-), organic phosphates (R_3PO_4), or esters of poly(ethylene glycol) $RCO_2CH_2(O(CH_2)_2O)_nCH_2O_2CR$. When used internally, 0.1–1.0% of the antistat is added to the plastics. Concentrations as high as 2% of antistat are applied to the plastic surface.

Antistatic agents are hydrophilic, that is, they attract moisture from the air. Thus, their effectiveness is dependent on a minimum relative humidity of 25%. Effectiveness is related to their hydrophilicity. It is assumed that under conditions of moderate humidity, a slightly conductive layer of water is formed on the surface of plastics containing antistatic agents. This layer tends to neutralize the static charge.

Lubricants such as zinc stearate, high molecular weight esters, and paraffin wax must be added to resins such as PVC and ABS to facilitate the processing of these materials. It is assumed that these additives reduce the coefficient of friction between plastic molecules and act as slip agents. These lubricants and other mold release agents such as talc are often dusted on mold surfaces to prevent adhesion of the plastic to the mold. Antiblocking agents such as fatty acids or poly(vinyl alcohol) are sometimes added to plastic sheet or film to prevent adhesion between layers of plastics such as polyethylene or poly(vinyl chloride).

Inorganic silicates or fumed silica are often added to plastics such as PVC to reduce surface adhesion or to produce a flattening effect. Finely-divided polyfluorocarbon particles in the 10 micron range are often added to plastics to assure release from metallic surfaces during processing and to provide plastics with a low coefficient of friction.

The surface of fillers and reinforcing agents such as fibrous glass are often treated with silane coupling agents in order to improve adhesion between these fillers and the resin. A typical coupling agent may be represented by the formula $((RO)_3SiCl)$. It is assumed that the chlorine is replaced by an active hydroxyl as a result of the reaction with water on the filler surface. Silicones are also added to polyurethane cellular plastics to lower the surface tension and thus stabilize the bubbles produced in the production of plastic foams.

BEHAVIORAL OBJECTIVES

Because of the presence of relatively weak bonds, it is customary to add stabilizers to PP and PVC. The stiffness of the latter is overcome by the addition of plasticizer. After reading this chapter, you should understand the following concepts:

1. Carbon atoms in polymers having only one hydrogen atom such as those in PP are particularly susceptible to degradation unless stabilizers are present.
2. Mixtures of stabilizers are often used since, when properly selected, the combination is more effective than the individual additives.
3. Because of the weakness of the carbon-chlorine bond, it is necessary to add stabilizers and sunscreening agents when plastics, such as PVC, are exposed to heat and sunlight.
4. Since they react with hydrogen chloride (HCl) to produce insoluble salts, compounds of heavy metals such as lead salts are used as stabilizers for PVC.
5. The most widely used sunscreening agents are compounds such as the 2-hydroxybenzophenones which are capable of forming a loose compound or chelate because of the attraction between the positively charged hydrogen atom and the unshared electron pair in oxygen in the carbonyl group ($C{=}O$).

6. Dioctyl phthalate is widely used as a plasticizer to increase the flexibility of PVC.

7. Substances such as mixtures of antimony chloride and organic chlorides are used as additives to improve the flame resistance of plastics.

8. The tendency for plastics to attract dust may be reduced by the addition of antistats which reduce the static charge on the surface of plastics.

9. Lubricants such as zinc stearate are added to aid the processing of plastics.

10. The adhesion of plastic sheets to each other may be prevented by the addition of incompatible antiblocking agents such as fumed silica.

GLOSSARY

Antistat: an additive which reduces the static charges on plastic surfaces.
Antioxidant: a stabilizer which retards polymer degradation.
Blocking: the undesired adhesion of plastic layers to each other during storage.
Carbonyl group: C=O.
Celluloid: cellulose nitrate plasticized by camphor.
Chelate: a compound produced by the attraction of a positively charged atom such as hydrogen to an atom such as oxygen which has a high electron density, i.e., unshared electron pairs.
Chromophoric group: color bearing group.
DOP: Dioctyl phthalate.
Flame retardants: additives which increase the ignition temperature of plastics.
Hindered phenol: one having substituents on the carbon atoms adjacent to the phenolic carbon atom.
hv: a photon = (Plank's constant) (frequency of radiation).
Hyatt, John W.: the inventor of celluloid which is usually considered to be the first manmade plastic.
Hydrophilic: attracts moisture.
Ignition temperature: the temperature at which plastics ignite spontaneously in the presence of flame.
Initiation: the first step in a chain reaction.
Macroradical: a high molecular weight radical.
Nitrocellulose: a trivial name for cellulose nitrate.
P: symbol for polymer.
Plate out: the migration of incompatible additives to the surface of a plastic.
Plasticizer: an additive which makes stiff plastics flexible.
Plasticizer efficiency: a measure of the relative effectiveness of a plasticizer.
Propagation: the chain carrying step in a chain reaction.

PPH: parts per hundred.
Saturated: containing the maximum amount of hydrogen atoms.
Secondary plasticizer: an extender for a primary plasticizer.
Synergists: two compounds which reinforce the effectiveness of each other.
Tertiary carbon free radical: a radical such as $\cdot C(CH_3)_3$ which lacks hydrogen atoms on one carbon atom.
Unsaturated: a compound which is deficient in hydrogen atoms.
UV: ultraviolet radiation.
Zipper reaction: a degradation reaction that continues spontaneously once it starts.

SUGGESTED QUESTIONS

1. Which is more stable—HDPE or PP?
2. Which is more stable—an unsaturated polymer such as rubber or a saturated polymer such as a copolymer of ethylene and propylene?
3. Which is more stable—PP containing 0.5 PPH of an antioxidant or PP containing 0.25 PPH each of the two different type antioxidants?
4. Why are heavy metal soaps effective stabilizers for PVC?
5. What intermediates are formed by sunscreening agents such as 2-hydroxybenzophenones in the presence of ultraviolet light?
6. What was the first manmade plastic?
7. What is the most widely used plasticizer?
8. Which is a more effective flame retardant, antimony oxide or a mixture of this and an organic chloride?
9. Which plastic will have the greater attraction for dust, PVC or LDPE?
10. What additives are used to facilitate plastic processing?

ANSWERS

1. HDPE, because it contains few tertiary carbon atoms.
2. The saturated copolymer.
3. Usually the latter because of synergism.
4. Because the metal ions react with the HCl, produced by decomposition of the PVC, to produce insoluble salts.
5. Chelates in which the phenolic hydrogen (OH) is attracted to an unshared electron pair in the carbonyl group (C=O).
6. Celluloid, i.e., plasticized cellulose nitrate.
7. DOP
8. The mixture.
9. LDPE because it is less polar, i.e., a poorer conductor of electricity.
10. Lubricants such as zinc stearate.

chapter 4

Introduction to Plastics Technology

4.1 MIXING OF RESINS AND ADDITIVES

Additives such as those discussed in Chaps. 2 and 3 are usually incorporated with the resin when the latter is a liquid or a finely-divided solid. Accordingly, many additives are readily incorporated by dissolving or dispersing them in compatible liquids. Thus, finely-divided poly(vinyl chloride) (PVC) and appropriately selected colorants and stabilizers are readily dispersed in liquid plasticizers to obtain plastisols which may be heated in molds to produce flexible plastic parts. Likewise, pigments and stabilizers when dispersed in water in the presence of surface active agents are readily admixed with aqueous dispersions of polymers, such as poly(vinyl acetate) (PVAC).

Solutions of initiators, such as benzoyl peroxide, polyamine curing agents, and compatible colorants, may be added to liquid unsaturated polyester resins, liquid epoxy resins, and to partially polymerized methyl methacrylate, respectively. The blending of solid additives with finely-divided solid thermoplastics is more complex, but this compounding step is readily accomplished with appropriate processing equipment as shown in Fig. 4-1. Some manufacturers do not add pigments in their own plants but farm out this step to companies specializing in the production of colored plastic composites erroneously called *compounds*. In contrast to the term *chemical compound*, which denotes a reaction product, the term *plastic compound* is used to indicate a uniform blend of resin and selected additives.

When subsequent production processing involves intensive mixing, as is the case with extrusion or injection molding, compounding may be accomplished by a simple blending of a finely-divided resin with powdered additives in a simple tumbling-type blender or a ribbon blender. However, the usual practice is to use the powder blend as a premix and to subject this blend to intensive mixing in a heavy duty mixer such as a "Banbury" or "Ko-Kneader" mixer.

The intensively-blended plastic compound may be extruded as strands in an auxiliary extruder. These strands are cooled in water and chopped by a rotating knife to produce pellets with relatively good uniformity. Color concentrates, which may be blended subsequently with unpigmented plastic, may be produced by adding a large excess of pigment in the intensive mixer. Compounded pellets are usually reblended before packaging to assure product uniformity in large lots.

Thermosetting resins, that is, polymers that crosslink or cure during thermal processing such as phenolic novolac resins, are first ground to fine

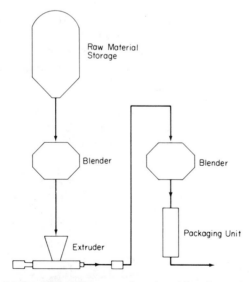

Fig. 4-1 Flow diagram showing blending of resin and additives.

powders and admixed with finely divided colorants, lubricants, fillers, and curing agents in a ribbon or tumbling blender. Mixtures such as those of *A* stage linear phenolic resins and additives are then polymerized somewhat or "advanced" to *B* stage resins and densified on preheated differential rolls or in an extruder. The product is cooled and chopped up to produce thermosetting resin molding powders.

Since fibrous reinforcing filters, such as alpha cellulose fibers or fibrous glass, are not readily mixed with powdered resin in a blender. It is customary to add these fillers to an aqueous syrup in a heavy duty mixer. The water is then removed by evaporation and the cooled dried product is pulverized. Other additives, such as lubricants, pigments, curing agents, and stabilizers, may be incorporated with the filled resin in a ball mill. The ball-milled compound is then densified on preheated differential rolls, cooled and ground, or comminuted.

4.2 CASTING: HOT MELTS, PLASTISOLS

The simplest type of casting is similar to the casting of metals except that a hot molten plastic, such as ethylcellulose or polyethylene, instead of the hot metal is poured in a mold and cooled. Since most plastics must be heated above their decomposition temperatures before they will flow under the force of gravity, the term *hot-melt compound* is usually used for the exceptions and the term *casting* reserved for describing a process in which polymerization occurs in a mold.

However, it should be noted that the rotational casting of powdered thermoplastic polymers such as polyethylene is also an important technique for the fabrication of large hollow items such as gasoline tanks. In this process, as shown in Fig. 4-2, a premeasured amount of powdered thermoplastic resin

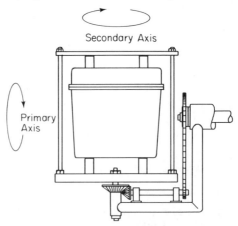

Fig. 4-2 A typical rotational molding process.

is placed in a closed mold which is then heated and rotated until the molten resin is distributed uniformly on the inner surface of the mold. The hot mold is then cooled with cold water and the cold cast part is removed.

The term casting is customarily applied to describe a simple process in which a mixture of a curing agent and low viscosity liquid resin is poured into a simple mold and polymerized at a higher temperature. The pioneer phenolic, polyester, epoxy, and poly(methyl methacrylate) plastic end-products were produced by a casting technique. A typical casting operation is illustrated in Fig. 4-3.

It is customary to partially polymerize a mixture of an initiator, such as benzoyl peroxide, and a monomer, such as methyl methacrylate, until a syrup or prepolymer is obtained. The syrup is then poured into a mold cavity and the polymerization is continued at elevated temperatures in the absence of oxygen. This casting process is used in the filling of teeth and in the making of dentures.

The term *potting* is used to describe the process in which the resin, such as an epoxy resin, impregnates or holds items, such as electrical assemblies, in place. *Encapsulating* describes a potting process in which the embedded parts are held in place by a cast cellular resin such as polyurethane. Documents and biological specimens are often embedded in a clear cast polymer of methyl methacrylate.

Fibrous glass-reinforced polyester and epoxy resin structures are made by the casting of a mixture of resin and curing agent in the presence of a fibrous glass reinforcing filler. A peroxy-type initiator is customarily used for curing polyester resins while a polyamine is often used for curing epoxy resins at ordinary temperatures. Metal-filled epoxy resins may be cast in inexpensive molds to produce forming tools used in the aircraft and automobile industries.

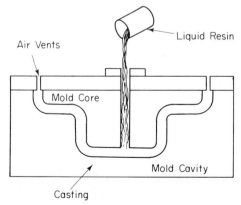

Fig. 4-3 Typical liquid process resin casting process.

When an emulsion of poly(vinyl chloride) is spray dried, the finely-divided resinous particles are coated with a thin layer of surface active agent. When these uniformly-sized particles are dispersed in a liquid plasticizer, such as dioctyl phthalate, a high viscosity resin paste is produced. More useful low viscosity pastes called *plastisols* are obtained when PVC particles of varying size are dispersed in a liquid plasticizer.

Articles such as rain boots or doll heads may be produced by slush or rotationally molding a compounded plastisol at 325°F. The compounded mixture contains pigments and stabilizers dispersed in the liquid dispersion of PVC and plasticizer. When the plastisol in contact with the hot mold surface has fused, the excess liquid plastisol is drained off and the adhered layer is heated to assure complete fusion of the plasticizer and resin.

A more uniform distribution of the plastisol in the finished part may be obtained by rotating the heated mold in three dimensions. Plastisol castings such as tool handles may be produced by dipping a form into a plastisol and heating the adhered resin at 325°F. The addition of bentonite clay increases the viscosity of plastisols. These so-called plastigels retain their viscosity when heated and thus relatively-thick sections may be obtained by the application of these clay-filled products.

4.3 CELLULAR PLASTICS

The modification of plastics by the incorporation of solids or fillers and the addition of liquid plasticizers has been discussed in previous sections. Polymers may also be modified by the addition of a gas to produce cellular plastics or foams. The original expanded polymer, which was called *sponge rubber*, was produced over 50 years ago by a process similar to that used for making biscuits and cakes. The blowing agent in the bakery products and the sponge rubber was sodium bicarbonate or baking soda which, as shown in Fig. 4-4, produces carbon dioxide when heated.

$$2NaHCO_3 \xrightarrow{\Delta} Na_2CO_3 + H_2O + CO_2 \uparrow$$

Fig. 4-4 Thermal decomposition of sodium bicarbonate.

Sponge rubber has been replaced to a large extent by foamed rubber. The latter is produced by the incorporation of a gas, such as air, into compounded rubber latex before curing. Latex foam has an open cell or multicellular structure in which the cells are interconnected. Other cellular products may have an open or closed cellular structure depending on the viscosity of the polymer at the time of blowing. Unicellular or closed-celled foams are produced when a gas is added to a polymer with high viscosity.

Foamed plastics can be produced by the use of either the rubber sponge or rubber foam technology, but most cellular plastics are produced by the addition of volatile liquids or chemical foaming agents. More than 600 million lbs of flexible polyurethane foam and over 300 million lbs of rigid polyurethane were produced in the USA in 1973. The annual production of rigid polystyrene foam in the USA is over 400 million lbs.

Polystyrene foam is being used to produce insulated beverage cups, meat trays, egg cartons, foundry casting molds, as concrete aggregate for Arctic highways and for furniture. Cellular cellulose acetate is used where good heat resistance and low water absorption properties are essential. Silicone foams, considered to be physiologically harmless, are characterized by excellent resistance to high and low temperatures and good electrical properties.

Flexible poly(vinyl chloride) foams are used as fabric coatings and for profiles in place of wooden structures. Plastic beads that expand simultaneously with the gas explosion of polyurethane yield a flexible foam with a high compression modulus, resulting in a dramatic improvement in cushion comfort.

As shown in Fig. 4-5, nitrogen is produced when azobisisobutyronitrile (AIBN) is heated. This gas, which is also obtained by the thermal decomposition of other azo compounds such as azobisformamide (ABFA), can also

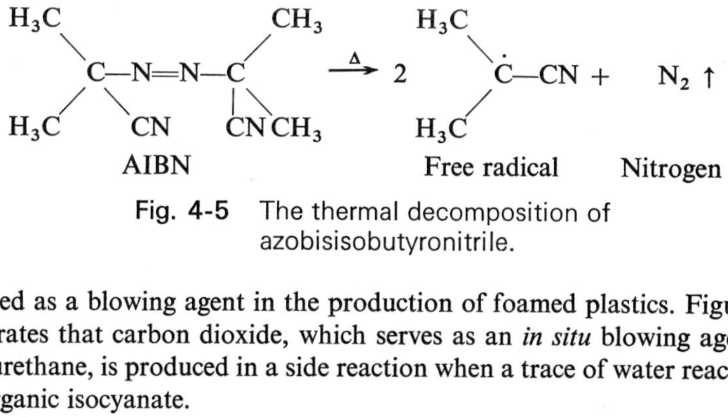

$$(CH_3)_2C(CN)-N=N-C(CN)(CH_3)_2 \xrightarrow{\Delta} 2\ (CH_3)_2\dot{C}-CN + N_2 \uparrow$$

AIBN 　　　　　　　　　　　Free radical　　Nitrogen

Fig. 4-5　The thermal decomposition of azobisisobutyronitrile.

be used as a blowing agent in the production of foamed plastics. Figure 4-6 illustrates that carbon dioxide, which serves as an *in situ* blowing agent for polyurethane, is produced in a side reaction when a trace of water reacts with an organic isocyanate.

A similar side reaction takes place when a trace of water is added to a mixture of a diisocyanate, such as tolylyl diisocyanate (TDI), and a

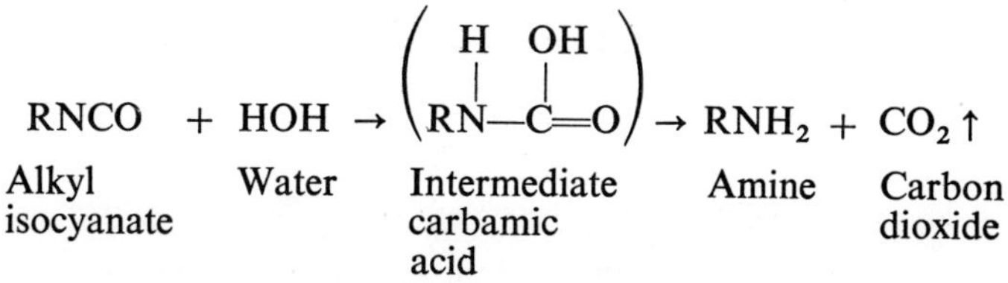

$$RNCO + HOH \rightarrow \left(RN(H)-C(OH)=O \right) \rightarrow RNH_2 + CO_2 \uparrow$$

Alkyl　　　Water　　Intermediate　　Amine　　Carbon
isocyanate　　　　　carbamic　　　　　　　　dioxide
　　　　　　　　　　acid

Fig. 4-6　The reaction of an alkyl isocyanate and water.

prepolymer such as poly(propylene glycol). The final product is flexible if there are many methylene groups (—CH_2—) or methylene ether groups (—CH_2OCH_2—) between the urethane $\left(\begin{array}{c} H \\ | \\ -O-C-N- \\ \| \\ O \end{array}\right)$ groups. The equation for the general reaction for the preparation of a polyurethane is given in Fig. 4-7. The process for making a flexible polyurethane foam is shown in Fig. 4-8.

$$n O{=}C{=}N(CH_2)_n N{=}C{=}O \; + \; n\, HO(CH_2)_n OH$$

   Organic diisocyanate   Poly(methylene glycol) ↓

$$\left(\begin{array}{c} H \qquad\quad H \\ | \qquad\qquad | \\ -C-N(CH_2)_n N-C-O(CH_2)_n O- \\ \| \qquad\qquad\quad \| \\ O \qquad\qquad O \end{array}\right)_n$$

Polyurethane

Fig. 4-7 Equation for the preparation of a polyurethane.

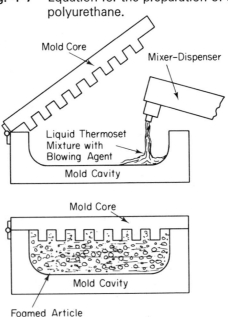

Fig. 4-8 Polyurethane foam production showing open and closed mold.

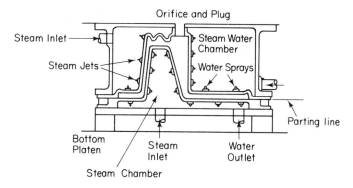

Fig. 4-9 Molding expandable polystyrene beads.

Expandable beads may be molded as shown in Fig. 4-9 and mixtures of polystyrene and pentane may be extruded to produce continuous cylinders or blocks. The latter may be cut or carved to produce almost any desired shape. When expandable beads are molded under pressure, a hard rigid exterior skin is formed. Moldings of this type are unusually strong and are used for making furniture and other structures.

Plastics with controlled densities, called *syntactic foams*, may also be produced by filling polymers with hollow spheres. The latter, called *microballoons*, may consist of glass, carbon, or phenolic resin. In addition to polyurethane and polystyrene, cellular plastics may be produced from ABS, cellulose acetate, ionomers, polyethylene, PVC, ureas, phenolics, and epoxy resins.

Deep water plastic floats are produced by incorporating silica microballoons into epoxy resins. Castable, flexible foams which are also useful for deep water service are obtained by incorporating glass microballoons in flexible urethane resins. Cellular products, called *multifoams*, are produced by the addition of gas blowing agents to syntactic foam compositions. Products with unusually high compressive strengths are obtained by using bubbles of different sizes in these multifoams.

Molded polyurethane foams that develop their own surface skins are obtained when these foams are produced in the presence of volatile solvents. These integral-skin foams which are being produced at a multimillion pound annual rate have great growth potential. The tough structural foams produced from other resins may be processed by many unique techniques.

Intricate shapes are constructed by assembly strips of structural foam produced by the expansion of an extruded foam by the Celuka method.

Self-skinning foams are produced in the Cincinnati Milacron process by feeding the foam into warm molds. Integral-skin foams are produced by injection molding of high-density polyurethane foams in the Duromer, Hoover, Isoderm, Rubicast and Union Carbide processes. The latter process, which produces a cellular core between two solid integral skins, may be used with many other resins.

4.4 REINFORCED PLASTICS AND LAMINATES

Glass *reinforced plastics* (GRP), which are made by the incorporation of glass fibers in various resins, are now produced in the USA at an annual rate in excess of 1 billion lbs. Over 80% of this volume consists of glass reinforced polyester resins. Considerable quantities of fibrous glass are also used to reinforce cellular plastics and thermoplastics.

The pioneer application of fibrous glass as a reinforcement for resins was for the production of aircraft radomes during World War II. Marine, ground transportation, construction, and corrosion-resistant applications now account for 75% of the end uses of fibrous-glass reinforced plastics. After cessation of hostilities, GRP was used for the construction of boat hulls, automobile bodies, corrugated building panels, and various structural parts.

The pioneer resin was obtained by the *in situ* free-radical polymerization of a styrene solution of an unsaturated polyester obtained by the condensation ethylene glycol and maleic anhydride. This composition was patented by Ellis in 1940. Subsequently, Muskat patented a composition in which maleic anhydride was replaced in part by phthalic anhydride, resulting in the type of polyester resin most widely used today.

Crude GRP structures may be produced by a hand layup technique which consists of a simple impregnation of a fibrous glass mat with a styrene solution of unsaturated resin containing benzoyl peroxide. This composite is placed on a mold and cured by heating the reinforced resin. A curable filled resin may also be prepared and applied by a spray-up technique in which compounded liquid resin and chopped glass fibers are passed simultaneously through a proprietary spray gun. Such mixtures may be used for built-up coatings on roofs and for patching.

The most reinforced plastics are produced by more sophisticated methods. For example, fibrous-glass reinforced polyester panels are produced by impregnating a layer of fibrous glass mat with compounded resin and repeating this process with enough layers until the desired thickness is obtained. This so-called *contact lay-up technique* may include an initial highly-filled resinous layer called a *gel coat*. Finely-divided thermoplastic resins may be added to the polyester resin in order to obtain a smooth, low profile, low shrink surface which may be readily painted.

Sheet molding compounds (SMC) and bulk molding compounds (BMC) are produced by compounding a mixture of polyester resin, fillers, thickeners, pigments and other additives with fibrous glass. BMC contain short ($\frac{1}{4}''$) fibers while the SMC consist of layers of longer filaments. As shown in Fig. 4-10 the impregnated fibrous glass sheets may be cured by heating between platens under low pressure. The smoothness of the surface of these cured sheets will depend on the smoothness of the metal platens and the presence of a gel coat of finely-divided thermoplastic resin.

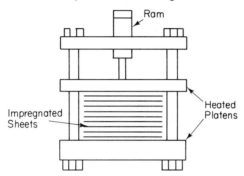

Fig. 4-10   Heat curing of GRP in a platen press. Premix or bulk molding compound may be cured by heating under pressure in a mold cavity.

Complex shaped articles may be made by heating sheet molding compound or bulk molding compound in matched metal molds. Because of the high initial cost of the molds, this sophisticated technique can be justified only for large volume runs. When only a few thousand parts are to be produced, less expensive heat conductive male molds are used. In this technique, the impregnated mat is draped over the mold and the assembly is protected by a nonadhesive film and held in place by an inflated rubber bag which maintains contact between the molded part and the mold while the assembly is heat cured.

Complex-shaped structures with unusually good strength may be produced by the filament winding process. As illustrated in Fig. 4-11, a strong complicated shaped structure such as a nose cone may be fabricated by passing a single fibrous glass strand or filament through a resin bath and winding the impregnated filament on a revolving mandrel with an appropriate design in accordance with a predetermined pattern. The finished article is obtained by removing the reinforced plastic article after oven curing. The physical properties of the article are dependent on the winding angle and the number of layers of filament used.

Continuous shapes such as fishing rods are produced by continuously

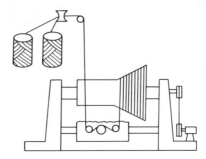

Fig. 4-11 Sketch of essential elements of filament winding process.

pulling a bundle of resin impregnated filaments through a die heated at temperatures of 250–300°F. The emerging cured rod may be cut to any desired length. This technique called *pultrusion* has been used to produce I beams, sheets, solid rods, and hollow structures such as pipe.

Approximately 90% of reinforced plastics are thermoset or crosslinked composites. However, reinforced thermoplastic composites are being produced at an annual rate in excess of 100 million lbs and are growing rapidly in importance. The use of thermosetting resins for composites is limited to polyester, epoxy, phenolic, diallyl phthalate and silicone resins but almost all the thermoplastics have been used as reinforced plastics (RTP). Fibrous glass has been used almost universally for reinforcing thermoplastics but asbestos has been used only to a limited extent. Potassium titanate crystals have shown promise for nylon reinforcement.

In general, the toughness, stiffness, resistance to stress cracking and creep, dimensional stability, impact resistance, tensile strength, compression strength, resistance to fatigue, and resistance to abrasion are increased substantially and mold shrinkage and the coefficient of expansion are reduced when 25–40% fibrous glass is added to a thermoplastic. This readily recognizable improvement counteracts the decrease in physical properties caused by the incorporation of nonreinforcing additives such as flame retardants and by foaming. Actually, foams produced from nylon, polystyrene, and polypropylene and reinforced with 20% fibrous glass have load bearing capacities and heat deflection temperatures superior to those of the unfilled solid resins.

Sheets of polypropylene or styrene-acrylonitrile copolymer filled with 40% fibrous glass may be formed on conventional metal stamping machinery. These reinforced thermoplastic sheets, called *azdel and stampglas*, are characterized by design flexibility and dimensional stability and can be formed at rates as fast as 360 parts per hour.

Both decorative and industrial laminates are reinforced plastic composites produced by superimposing resin-impregnated layers of paper or fabric, removing any solvent present, and curing the assembled sheets under heat and pressure. Decorative laminates used for walls and countertops usually consist of a core made up of several sheets of phenolic resin-impregnated kraft paper sheets topped by a melamine resin-impregnated "decor" paper sheet. The latter is protected from surface damage caused by solvents, detergents, and heat by use of an overlay of melamine resin-impregnated alpha cellulose tissue.

The entire assembly is molded in a multi-opening laminating press at about 275°F and 1,500 psi. Presses holding as many as 10 sheets per opening are available for molding sheets as large as 5 ft by 12 ft. The decorative pattern that characterizes the finished laminated sheet may be applied to the cellulose tissue by hand sketching, silk screening, letter press, or rotogravure printing.

Industrial laminates are available in many different standard grades. They are similar to the core stock of the decorative laminate. The impregnated sheet may consist of paper, asbestos or fibrous glass mat, or fabric. Phenolic, melamine, polyester, epoxy, or silicone resins may be used as impregnants in accordance with the ultimate end-use of the laminate. Sheet laminates are produced by pressing in a multi-opening laminating press. Tubes or rods may be produced by rolling sheet stock, pultrusion, or convoluted winding.

The processes discussed in this chapter are dependent to some extent on plastic flow. The latter which is important for molding and extrusion processes is discussed in Sec. 4.5.

4.5 INTRODUCTION TO RHEOLOGY

The branch of science concerned with the study of deformation and flow of matter is called *rheology*. The rheological behavior of solids and fluids are described by two extremely different branches of mechanics called *solid mechanics* and *fluid mechanics*. However, these two extemes are of limited interest to the plastics technologist who is concerned primarily with materials that exhibit both solid and fluid properties. These products are said to be *viscoelastic*.

An elastic solid may be subjected to a shearing stress, that is a tangential force like that used in the wiping of grease on a pie tin. Under these conditions of stress, the solid is deformed temporarily but returns to its original shape instantaneously after the stress is removed. Thus, it may be said that the energy consumed in the deformation of the elastic solid is recoverable when the stress is removed in this reversible process. In contrast, a viscous liquid undergoes continuous deformation when stress is applied and this laminar flow is irreversible.

All plastics are viscoelastic, and their behavior under stress is much more complex than that described above for an elastic solid or a viscous liquid. Plastic technologists must consider the rheological properties of polymeric solutions, dispersions, melts and solids but only the last two are of interest for most plastic processing.

When an amorphous plastic is subjected to stress, the carbon-carbon bonds in polymer backbone are stretched or compressed, and the normal tetrahedral bond angles are distorted. If the instantaneously-applied shearing stress (S) is small, it is proportional to the elastic deformation or strain (γ). The proportionality constant (G) is the shear modulus of elasticity. Since this relationship was expressed first by Hooke in 1876, this completely elastic behavior is said to be *hookean*.

According to Hooke's law, the stress is proportional to the shear strain. Hookean behavior is independent of time. It is similar to that of a stretched elastic spring whose modulus is G. This behavior, which is typical of many plastics below the glass transition temperature (T_g), is described by Hooke's law for shear as shown in Fig. 4-12.

$$\frac{\text{Shear stress } (S)}{\text{Shear strain } (\gamma)} = \text{Shear modulus of elasticity } (G)$$

or

$$S = G\gamma$$

Fig. 4-12 Hooke's Law.

When one considers tensile deformation, it is necessary to introduce a term called the *elastic shear compliance J* which is equal to $1/G$ and to consider an elastic constant (E) which is related to Poisson's ratio (μ) by the expression shown in Fig. 4-13. The values for Poisson's ratio vary from 0.25 to 0.5 and

$$E = 2(1 + \mu)G$$

Fig. 4-13 The relation of the elastic constant E to the shear modulus (G) and Poisson's ratio (μ).

are in the order of increasing elasticity. Thus, typical μ values for crystalline, glassy and elastic materials would be 0.25, 0.33, and 0.5 respectively. The μ value for PVC will increase as one increases the plasticizer content and that of natural rubber will decrease as one increases the sulfur content in the compounding recipe.

Amorphous plastics also exhibit some irreversible flow similar to that described by Newton in 1685. According to this concept, a completely viscous or newtonian fluid is one in which the shear stress (S) is proportional to the rate of change of velocity with distance. This rate is called the velocity gradient $d\gamma/dt$ and the proportionality constant for this viscous flow relationship is the coefficient of viscosity (η). Newtonian behavior is similar to that of a dash pot consisting of a piston and a cylinder containing a viscous liquid with a viscosity of η. Newton's law for viscous flow is described by the expression shown in Fig. 4-14.

$$\text{Shear stress } (S) = \left(\text{coefficient of viscosity }(\eta)\right)\left(\text{velocity gradient }\left(\frac{d\gamma}{dt}\right)\right)$$

$$S = \eta \left(\frac{d\gamma}{dt}\right)$$

Fig. 4-14 Newton's law.

Molten polymers flowing under extremely-low shear rates will flow like a highly viscous newtonian fluid but will exhibit the hookean behavior characteristic of a highly elastic solid if the shear rates are sufficiently high. Thus, the actual rheological behavior of most plastics are much more complex than that described by either Hooke's or Newton's law. Actually, one must consider the behavior of viscoelastic materials as described by a spectrum in which the ideal elastic and pure viscous responses are the two extremes of the spectrum.

When a linear amorphous plastic is stressed at a temperature above T_g, it exhibits a retarded elastic response or relaxation as a result of a disentanglement and uncoiling of polymer chains. The extent of recovery after the application of this stress is related inversely to the length of time in which this stress is maintained. If the stress is relieved after a fraction of a second, the behavior may approach that of an elastic solid and the plastic is said to have a good memory. Whereas if the stress is maintained on a plastic for a long interval of time, its behavior will approach that of a purely viscous fluid and it will have little if any memory. The flow occurring after the application of stress for long periods of time is called *creep* and is related to disentanglement and mutual slippage of polymer chains.

The viscoelastic behavior of a plastic may be represented in a crude way by combinations of the spring and dash pot models used to describe hookean and newtonian behavior. When these models are placed in series, the model which is called the *Maxwell model* may be described by the rheological equation shown in Fig. 4-15. The creep for the Maxwell model consists of an

$$\frac{d\gamma}{dt} = \frac{dS}{dt}\left(\frac{1}{G}\right) + \frac{1}{\eta}S$$

Fig. 4-15 Equation for the Maxwell model.

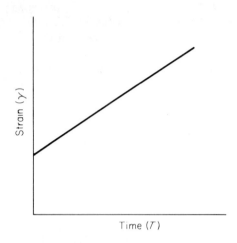

Fig. 4-16 Creep curve for Maxwell model.

initial elastic response by the spring followed by a slow viscous flow in the dash pot as illustrated in Fig. 4-16.

The viscoelastic behavior of a plastic may also be represented in a crude way by a Voigt model in which the spring and dash pot models are parallel. This model is described by the equation in Fig. 4-17. The Voigt model

$$S = G\gamma + \eta \frac{d\gamma}{dt}$$

Fig. 4-17 Equation for the Voigt model.

exhibits retarded elastic deformation in creep and retarded elastic recovery in recovery from creep. The total stress in this model is absorbed by the dash pot which then transfers its load to the spring which bears the total stress at equilibrium. This behavior is represented by the deformation curve shown in Fig. 4-18.

A more representative but still crude model of viscoelastic behavior is obtained by placing the Maxwell and Voigt models in series. Hysteresis or the energy associated with irreversible deformation which is dissipated as heat and the different paths for retraction and extension are exhibited by this complex model. However, still more complex models are required for a truer representation of the behavior of plastics under stress. It is also important to note that viscoelastic flow can be investigated independently of the elastic behavior.

The flow of low molecular weight liquids may be linear or newtonian, but this flow for plastics is nonlinear since it is impeded by entanglements of the polymer chain when the degree of polymerization ($\overline{PD}$) is greater than

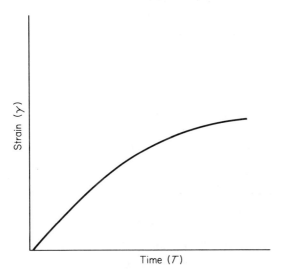

Fig. 4-18 Creep curve for the Voigt model.

600. Thus, for plastics, the melt viscosity (η) is proportional to the weight average molecular weight ($\overline{Mw}$) raised to an appropriate power. As depicted in Fig. 4-19, an exponent of 3.4 in melt viscosity equations is valid for many high molecular weight plastics.

$$\eta = k\overline{Mw}^{3.4}$$

Fig. 4-19 Relationship of melt viscosity and molecular weight.

In spite of this tendency to flow when stressed, the behavior of most plastics resembles that of elastic solid until a threshold stress (S_0) is exceeded. When the stress exceeds this critical value, the plastic flows like a viscous fluid. Plastics that do not flow until the stress exceeds such a yield stress value are called *Bingham* or *ideal plastics*. The Bingham relationship is shown in Fig. 4-20.

$$S - S_0 = \eta \frac{d\gamma}{dt}$$

Fig. 4-20 Flow for a Bingham plastic.

Obviously, the flow behavior of a typical plastic is much more complex than that described by the Bingham model. A more appropriate relationship is described by a general equation called the *Herschel-Bulkley equation* in which

ϕ is a function of the coefficient of viscosity (η) and the exponent n may have any value. The Herschel-Bulkley equation shown in Fig. 4-21 is identical to the Bingham relationship when $\phi = \eta$ and $n = 1$. When S_0 is 0, it is identical with Newton's equation.

$$(S - S_0)^n = \phi \frac{d\gamma}{dt}$$

Fig. 4-21 Herschel-Bulkley equation.

The rate of flow of plastics increases as the temperature is increased. The relationship of the viscosity (η) to temperature in degrees kelvin (T) is shown by the Arrhenius equation in Fig. 4-22. In this equation, R is the gas constant, E is the activation energy for flow and A is a constant.

$$\eta = Ae^{E/RT}$$

Fig. 4-22 The relationship of viscosity to temperature.

If one takes the logarithm of both sides of the Arrhenius equation, one arrives at the expression in Fig. 4-23 which states that small increases in temperature, such as 10°C, produce large decreases in viscosity. According to

$$\log \eta = \log A + \frac{E}{2.3RT}$$

Fig. 4-23 Logarithmic form of the Arrhenius equation.

this equation, a straight line with a slope of $E/2.3R$ should be obtained when one plots $\log \eta$ against the reciprocal of the temperature. The intercept of this line with the vertical axis would be $\log A$, as shown in Fig. 4-24.

Since a knowledge of specific flow properties is essential for the processing of plastics under optimum conditions, several empirical methods have been devised to measure flow. In one commercial torque rheometer, called the *Brabender Plasticorder*, the shear rate and temperature are varied and thus processibility data are provided in terms of shear, torque, and temperature.

Flow data for thermosetting plastics such as phenolic plastics may also be obtained by measuring the time in seconds for flash to form and to cease forming in a small mold at a temperature of 103°C and a pressure of 20,000 psi. The melt flow index is used to measure the flow of thermoplastics such as polyethylene and polypropylene. This parameter, which is usually called simply *melt index*, is the weight in grams of plastic extruded in 10 minutes at

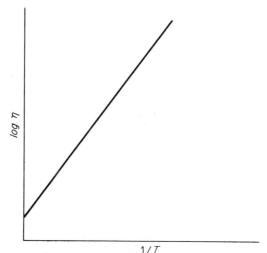

Fig. 4-24 The effect of temperature on viscosity.

43.2 psi at a specified temperature. Temperatures of 190°C and 230°C are specified for polyethylene and polypropylene, respectively.

BEHAVIORAL OBJECTIVES

Since uncompounded plastics are seldom used, the mixing of resins and additives, which is called compounding, and the processing of these so-called plastic compounds to produce useful articles is of major importance.

After reading this chapter you should understand the following:

1. The term plastic compound is used to denote a uniform blend of resin and appropriate additives. It is customary to blend these ingredients in a ribbon blender and then to blend the premix intensively in heavy duty equipment.

2. Either hot thermoplastics or liquid thermosets may be cast in a mold to produce a finished article after the molten plastic cools or the cold liquid resin crosslinks. Fibrous glass reinforced articles are formed by a casting process.

3. Flexible articles may be cast by heating plastisols in a simple mold. Plastisols consist of a suspension of finely divided unplasticized hard resin in a liquid plasticizer. The latter diffuses into the resin particles to produce a uniform plasticized product when heated above 325°F.

4. Cellular plastics or foams are produced by injecting gas into a viscous plastic before solidifying the cellular mixture.

5. A gaseous propellant, carbon dioxide, may be formed *in situ* when polyurethanes are produced in the presence of traces of water.

6. Light weight solid articles, such as furniture, can be produced by the formation of cellular products in the presence of solvents in appropriate molds. There are many proprietary techniques available for the production of these integral skin foams.

7. Reinforced polyester articles are produced by the room temperature-casting of a resin-impregnated fibrous glass composite, the heat forming and curing of an impregnated sheet (SMC), or the molding of a mixture of short glass fibers and resin (BMC).

8. The strength and heat resistance of crystalline thermoplastics are improved by reinforcing with fibrous fillers.

9. Decorative laminates are produced by laying up resin-impregnated sheets and using a melamine resin impregnated decorated sheet for the top layer. The entire composite is then heat-cured in a laminating press.

10. Since plastics behave both as elastic solids and viscous liquids, they are said to be viscoelastic.

11. An ideal elastic solid obeys Hooke's law, i.e., the applied stress (S) is proportional to the strain (γ).

12. Poisson's ratio is about 0.25 for stiff solids and increases to 0.5 for flexible plastics.

13. A disentanglement and uncoiling of polymer chains occurs when an amorphous plastic is subjected to stress.

14. The creep or cold flow of a plastic is related to a combination of elastic and viscous properties.

15. The melt viscosity (η) of high molecular weight polymers is proportional to the molecular weight raised to the 3.4 power.

16. Plastics having properties described by the Bingham equation resist deformation until the stress (S) exceeds a critical value (S_0).

17. The viscosities of molten plastics decrease exponentially with temperature in accordance with the Arrhenius equation: $\eta = Ae^{E/RT}$.

GLOSSARY

Advancing: partially curing by mechanically working and heating of a plastic compound.

AIBN: azobisisobutryonitrile: a compound that produces nitrogen gas and free radicals when heated.

Arrhenius equation: $\eta = Ae^{E/RT}$.

Bingham equation: $S - S_0 = \eta(d\gamma/dt)$.

GLOSSARY

Bingham plastic: a plastic that resists flow until the stress exceeds a critical value (S_0).
BMC: bulk molding compound, usually a fibrous glass-unsaturated polyester composite.
Casting: the solidification of a liquid resinous composition in a mold without the use of pressure.
Creep: cold flow.
Dash pot: a model consisting of a piston in a cylinder containing a viscous liquid.
dγ/dt: velocity gradient or flow rate.
e: symbol for base of natural or naperian logarithms = 2.718.
E: an elastic constant related to Poisson's ratio (μ).
η: coefficient of viscosity.
Foam: a mixture of a gas and a solid
G: symbol for shear modulus.
γ: symbol for deformation or strain.
Gel coat: a viscous filler-free surface layer.
Herschel-Bulkley equation: $(S - S_0)^n = \phi(d\gamma/dt)$.
Hooke's law: $S = G\gamma$.
Laminate: a composite consisting of layers of reinforcing members and resin.
Log: logarithms based on the number 10 which are called common logarithms.
Maxwell model: a model representing viscoelastic deformation.
Melt index: a measure of flow. The weight in grams of polyethylene extruded through an orifice in 10 minutes at a specified temperature and pressure.
Multicellular: open cell or interconnecting cells.
Newton's law: $S = \eta(d\gamma)/(dt)$.
Plastigel: a very viscous plastisol.
Plastisol: a suspension of finely divided unplasticized resin such as PVC in a liquid plasticizer.
Potting: the casting of a hardenable resin in a simple mold usually with some object such as a wire embedded in the viscous resin.
Premix: a mixture of resin and appropriate ingredients used to produce a so-called plastic molding compound.
Pultrusion: a process in which resin impregnated filaments are pulled through a die before curing.
PUR: polyurethane.
Rheology: the science of flow.
S: symbol for applied stress.
V: symbol for elastic shear compliance.
Viscoelastic: having both viscous and elastic properties.
Viscosity: resistance to flow.
Voigt model: a model representing viscoelastic deformation.

SUGGESTED QUESTIONS

1. What are the essential ingredients in a plastic molding compound?
2. How would you produce a paper weight consisting of a coin or medal embedded in a clear plastic?
3. How would you set up a simple process for coating tool handles with a resilient nonconducting plastic?
4. What would be the simplest technique for producing a low density casting?
5. How would you produce a multicellular foam?
6. What causes the production of gas when a polyurethane resin is cured?
7. What is the simplest technique for producing a relatively strong vessel with a complex shape?
8. Why is a dark resin such as PF used in the interior layers of a decorative laminate?
9. Which will have a higher Poisson's ratio, plasticized PVC or PF?
10. What is the relationship of melt viscosity (η) to the average molecular weight of a plastic?
11. Why is it advantageous to select plastics which have properties described by the Bingham equation?
12. Which HDPE will have the higher viscosity, one with a low or high melt index.

ANSWERS

1. The resin, any required curing agent, lubricants and pigments as well as other additives such as fillers and stabilizers.
2. Advance methyl methacrylate to a viscous syrup, place the syrup and medal in a suitable mold and heat the composite slowly in order to complete the curing process.
3. Dip the tool handle in a viscous compounded plastisol and heat at 325°F. The thickness may be controlled by adjusting the viscosity, or by repeating the process.
4. Heat a uniform mixture of sodium bicarbonate and casting resin in an appropriate mold similar to the baking of baking powder biscuits.
5. By injecting the gas in a low viscosity resin before hardening or curing.
6. Carbon dioxide which is produced by a side reaction between water and the diisocyanate reactant.
7. Place layers of compounded unsaturated polyester or epoxy resin-impregnated fibrous glass mat over a wooden or plaster of Paris form and heat-cure the composite.
8. Principally for economical reasons.

9. Plasticized PVC, μ approaches 0.5, whereas the value for μ for PF approaches 0.25.
10. $\eta = \overline{Mw}^{3.4}$.
11. Bingham plastics can be processed by applying a stress greater than the yield stress (S_0), but they will not cold flow in service unless the threshold stress value is exceeded.
12. HDPE with a low melt index value.

chapter 5

Molding, Extrusion, and Calendering

5.1 COMPRESSION MOLDING

Thermoplastic molding powders may be placed in the lower section of a heated cavity and the mold similar to that shown in Fig. 5-1 may be closed under pressure. Since it is necessary to cool the mold before removing thermoplastic molded parts, this technique is not economical. However, compression molding is a classical procedure that has been used for almost 75 years for molding thermosetting plastics. It is more convenient than the molding of thermoplastics in that molded thermosetting plastic parts may be removed readily from a hot mold. The original presses were manually operated and hence entailed high labor costs, but the basic molding process has been improved by automation so that labor costs are competitive with the injection molding of thermoplastics.

When molding thermosetting plastics, it is customary to place a slight excess of the amount of molding powder required to fill the hot lower section

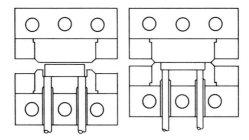

Fig. 5-1 Crude flash mold in open position on left and closed position on right.

of a mold cavity similar to that shown in Fig. 5-1. A thermosetting powder consists of partly-cured resin, fillers, pigments, mold release agents, and curing agents. It may be supplied to the mold cavity as a preformed tablet containing enough molding powder so that it will more than fill the space in the cavity.

In the molding operation, the upper section of the mold is closed down on the lower section, as illustrated by the left side of Fig. 5-1. Trapped air and gas escape and excess plastic exudes as flash while heat and pressure are maintained until the plastics is cured. The pressure is then released and the molded part is removed.

There are three general types of molds that vary with the amount of flash that is exuded. The flash mold, shown in Fig. 5-1, is the simplest type and is suitable for molding flat pieces. For more complicated parts, a semipositive mold is necessary in which the upper section or plunger has a tighter fit and its downward travel is limited by a sort of ledge on the bottom section, called a *land* or *stop*. Because of the tighter fit, less flash is produced in a *semipositive mold*.

The land or stop is eliminated in the *positive mold*. No flash is produced in mold and the downward travel of the plunger or force is controlled by the amount of molding powder in the mold cavity. Thus, the thickness of the molded part in the positive mold depends on the amount of molding powder used.

Molds are usually chromium plated. They may have multicavities and may be assembled from two or more parts. Molds may be steam heated but the use of electric heating coils or an electric heating cartridge is preferred. The term *tool* is used to describe the part of the mold that controls the size and shape of the molded part.

Press ratings are a measure of the total force a press can withstand. For example, a mold with an area of 25 sq in. in a 100 ton press will provide a pressure of 4 tons/sq in. A pressure of 3,000 lbs/sq in. is required to mold parts up to 1 in. in thickness. An additional 700 lbs/sq in. must be provided for each additional inch of thickness.

It is customary to heat the mold to 320–380°F when using cold general purpose (GP) phenolic molding powder. The curing time may vary from 30 seconds to a few minutes and may be shortened by increasing the temperature. However, the curing time must be sufficiently long to assure flow of the molten material to all parts of the mold cavity. The residence time of the molding powder in the mold may be reduced by preheating the preformed tablet for a short time interval at a temperature below the curing temperature, such as 290°F.

Labor costs and the time for compression molding may be reduced still further by a process called transfer molding. As depicted in Fig. 5-2, the hot preformed tablet is placed in a transfer pot or antichamber and forced through an orifice into the hot mold cavity.

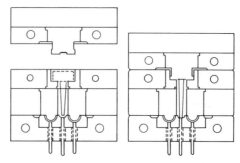

Fig. 5-2 A transfer mold showing heated preformed tablet in antichamber on left and after it has been forced through orifice to closed mold cavities on right.

It is customary to use multicavity molds in the transfer process and to discard the excess feed material outside the mold cavity that cures because of frictional heat. This waste consists of the principal material in the feed orifice called the sprue or stalk, and runners formed by curing the material between the sprue and the gates to the mold cavities. The latter have small cross-sections and thus the molded gates are readily broken off from the molded parts. Because of the fluidity of the plastic material, mold erosion is reduced in the transfer molding process.

Because of the high resistance to flow resulting from the sprue, runners, and gates, pressures as high as 25,000 lbs/sq in. are commonly used for transfer molding. The improved economy resulting from the shortened molding cycle in transfer molding more than compensates for cost of the material discarded as sprue, runners, and gates.

The two mold halves are often pried apart by hand in manually-operated compression molding. In semiautomatic molding processes, such as transfer molding, the molded parts are pushed free from the mold cavity by use of ejector pins that are mounted on an ejector pin plate. A combing device, which removes the molded parts from the ejector pins and deposits them away from the mold area, is used with automated presses.

The most highly-automated presses for molding thermosetting plastics are the screw transfer and injection molding presses. In the former process, the molding powder is preheated up to temperatures as high as 290°F as a result of preplasticization by a screw. This process permits the curing of parts with thick sections which is not possible with less automated processes. The fully-automated screw transfer molding process also reduces cure times so that the output per mold cavity is increased as much as 400%. Injection molding will be discussed in Sec. 5-4.

5.2 EXTRUSION

Extrusion is an unusually versatile process in which a granulated thermoplastic is converted continuously to a wide variety of continuous forms such as filament, rod, profile, sheet, film, inner tubes, and pipe. This process is also used for making parisons for blow molding, for coating different substrates such as paper or metal foil, and for providing insulating jackets for wire. The extrusion process may be used for compounding thermoplastics and thermosetting plastics and for supplying molten polymer in the injection molding process. More thermoplastics are processed by extrusion than by any other type of processing. Over eight billion pounds of thermoplastics were extruded in 1973.

As shown in Fig. 5-3, the extruder consists of a rotating archimedean

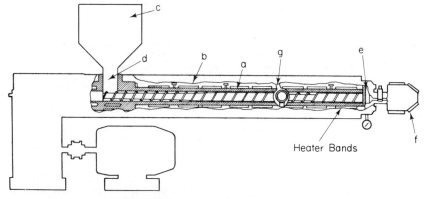

Fig. 5-3 Diagram of a single-screw extruder. Legend for numbers are given in the text.

screw labelled (a), enclosed in a heated cylinder or barrel labelled (b). The barrel houses the screw and provides a means for heating and cooling the extrudate in various forms as it advances from the hopper (c) through the feed throat (d) into the hot barrel and through the breaker plate and screen pack (e) to the die (f). The screen pack consists of several woven wire screens arranged so that the extrudate passes through the coarser screens before going through the finer areas. The breaker plate consists of a sturdy metal plate with many large holes. It supports the screen pack which removes any nonextrudable contaminants and improves the homogeneity of the extrudate by providing more intensive mixing.

Extruders with multiple screws are available but, because of its relative simplicity and utility, the single screw extruder is usually preferred (see Fig. 5-4). Twin screw extruders are better able to provide controlled rates of shear. They are used for compounding and extruding heat and shear-sensitive plastics such as rigid PVC. An output of 250 to 1,000 lbs/hr is possible on multiple screw extruders. Minimum shear is provided by the use of reciprocating rotating screw extruders. Two extruders linked to a common die are used for the coextrusion of two different plastics.

The extrusion process may be divided into zones. The first zone is the feed zone which starts just below the feed throat. This zone, in which initial mixing and plasticization occurs, is also called the *transport zone*. The plastic granules from the hopper are softened and melted by frictional heat and contact with the barrel wall as they are conveyed from the feed zone to the intermediate transition zone.

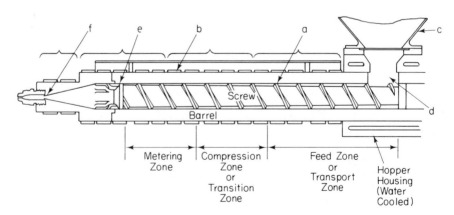

Fig. 5-4 Sketch showing details of screw and extruder zones.

The polymer melt is moved by the pumping action of the helically-flighted rotating screw through the metering zone. The molten material is then forced through the breaker plate and screen pack and through the die which changes the shape of the extrudate to the desired form. The flat dies used for sheet are available in widths up to 130 in. It is essential that uniform pressure be maintained across the entire channel of the die in order to produce sheet with uniform thickness. Flat dies are also used for coating paper or metal sheet.

Circular dies are used for extruding round rod, pipe, wire coating, or blown film. It is customary to insert a center mandrel and a spreader plate after the screen pack in order to assure uniform flow of the extrudate to the circular die used for blown film. These devices must be held in place in the barrel by spider arms and the passage of the extrudate around these arms causes the appearance of weld lines on the final extrudate. It is customary to seal the end of the hot extrudate with nip rolls while using air to expand the smaller plastic tube. Sheet film is obtained by slitting plastic tube.

Screws are characterized by the ratio of their length (L) to the inside diameter of the barrel (D) and the compression ratio. Barrels with standard diameters of $1\frac{1}{2}$ in.–$4\frac{1}{2}$ in. increments, 6 in. and 8 in. are available and barrels with larger diameters can be obtained by special order. For convenience, D is assigned a value of 1 and extruders with L/D ratios of 20:1, 24:1, 28:1, and larger are available by special order. Since high L/D ratios provide higher output and better melt uniformity, the trend is toward higher L/D ratios.

The effect of L/D ratios on extruder capacity is shown in Fig. 5-5. The screw consists of a cylinder root around which a single metal ridge or flight is spirally wrapped. The distance between repeating ridges or flights is called

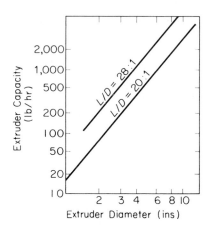

Fig. 5-5 Relationship of L/D ratio to extruder capacity.

the *lead*. The difference between the outside diameter of the screw and that of the cylinder root is called the *flight depth* and the helical angle is called the *pitch*. The latter is usually in the range of 12–20°.

It is customary to decrease the flight depth progressively as the screw goes from the feed to the metering zone. This provides for compression of the molten material. The ratio of the volume of one channel flight of the screw at the hopper end to that of one flight at the die end is called the *compression ratio*. The latter is usually 2:1 but may be as high as 4:1.

The root is usually hollow so that it may be water cooled. Extruders may be equipped with a vent (g) for the degassing of volatile material. The drive motor is usually connected to a gear reducer so that the screw may be turned at constant torque over a wide range of speed. The speed of the rotating screw may vary from 10 to 50 RPM but is held constant during any extrusion run. In general, the power consumption is independent of the speed on any specific extruder. The temperature of the cylinder may range from 100–400°C and is usually lower in multiple screw extruders.

Complex profiles are produced using appropriately-designed dies. Cellular extrudates are produced by expansion of the extrudate by the incorporation of low boiling liquids, such as pentane, or unstable compounds, such as ammonium carbonate, in the feed stock. Some of the basic shapes of the extruded foam are baseboards, corner strips, casings, doorstop strips, frames, shutters, and panels.

Over one billion pounds of plastics are extruded annually to produce plastic pipe. The principal polymers used as pipe are poly(vinyl chloride) (PVC), high and low density polyethylene (HDPE, LDPE), acrylonitrile-butadiene-styrene copolymer and polystryene-rubber blends. Polypropylene nylon, polyformaldehyde (acetal) and cellulose acetate butyrate are used to a lesser extent. The principal use of plastic pipe is for potable water distribution lines, irrigation pipe, drain, waste and vent piping (DWV), gas distribution lines, corrosion resistant pipe and electrical and telephone conduits, domestic plumbing, and sewer pipe.

Polyethylene pipe in sizes up to 48 in. in diameter is available. All types of thermoplastic pipe are used for renovating faulty metal pipe by inserting the plastic pipe and pulling it through the existing pipe line. In some instances, the thermoplastic pipe has been strengthened by wrapping with fibrous glass impregnated with epoxy resin. Chlorinated poly(vinyl chloride) (PVDV) pipe is used for hot water pipe.

Over two billion pounds of low density polyethylene film (LDPE) was extruded as film in 1973. Approximately 25% of this production was used to produce trash bags. Other plastic film bags are used for food and industrial packaging. Film in widths up to 40 ft is used for wrapping, tarpaulins, and as shrink pallets.

5.3 BLOW MOLDING

Hollow articles such as bottles are readily produced by blow molding, which is an adaptation of the glass bottle production technique. This process consists of placing a heat softened plastic tube or parison in a cold two-piece mold, heating the parison after closing its end, and inflating it with compressed air so that it takes the shape of the mold. The parison may be produced continuously by extrusion, cutting to proper length, and transferring to the mold for blowing.

The general scheme for extrusion blow molding is shown in Fig. 5-6.

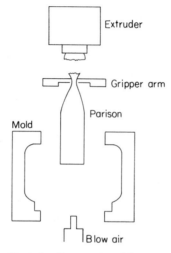

Fig. 5-6   Sketch of extrusion blow molding.

It is customary to cool the mold with cold water or liquid carbon dioxide in order to increase the number of cycles per minute. The various continuous processes include mounting of multiple molds using such designs as ferris wheels, rotary tables, and in-line mold over mold. In addition to the production of bottles of all sizes, the blow molding process is used to produce toys, tanks, household items, plumbing supplies, and automotive parts. Over a billion pounds of thermoplastics were used for blow molding in 1973.

5.4 INJECTION MOLDING

As shown in Fig. 5-7, some of the features of extrusion and blow molding are characteristic of the modern injection molding process. Because of many improvements in this versatile press, over 5 billion pounds of thermoplastics

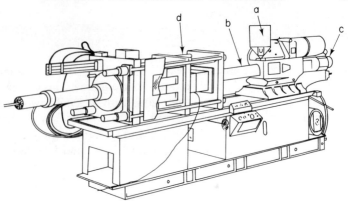

Fig. 5-7 Cross-section of an injection molding press.

and over 50 million pounds of thermosetting plastics were injection molded in 1973. Equipment with pressures up to 700 tons and with 16 cavity molds are now available. As a result of its inherent versatility and many improvements, injection molding is one of the fastest growing processing techniques used by the plastics industry.

Figure 5-7 illustrates that a measured volume of molding powder is conveyed by gravity feed from the hopper (a) to a heated cylinder (b). In the classical process shown, the molding power is heated and plasticized or plasticated in this cylinder and the molten plastic is forced by a reciprocating plunger (c) past a spreader or torpedo into a cooled and closed two piece mold (d). The cooled molded plastic part is then ejected when the mold opens. It then closes as the plunger begins its forword stroke in this cyclic mechanical process.

In a modification of this process, the single-stage plunger is replaced by a two-stage plunger. A larger quantity of molten plastic than needed for the mold cavity is conveyed to a second chamber, called a *shot chamber*, and a precisely-measured amount of this preplasticated product is conveyed to the mold cavity. Because of the longer residence time in the heated cylinder, the chances of injecting unplasticated material to the mold cavity are essentially eliminated.

In another modification, the extruder system is incorporated and a fixed screw is used to preplasticate the plastic and convey it to the shot chamber before injection into the mold cavity. In a more recent improvement, an in-line reciprocating screw is used to serve as both a plasticator and a ram. The design of this screw is similar to that shown in Fig. 5-4.

The parts of the mold may be held together by hydraulic pressure, by a mechanical or toggle device, or by a combination of these systems. A typical multicavity mold consists of a sprue or stalk which accepts the molten plastic and conveys it under pressure along passages called *runners* and then through constricted openings called *gates* to the mold cavities. Thus, when the mold is opened after cooling, the ejected plastic piece includes the molded parts as well as the runners and sprue. The mold is designed so that the parts are readily broken off from the runner at the gate.

When thermoplastics are molded, the sprue and runners, which account for about 5% of the charge, may be ground and remolded. However, this material must be scrapped when thermosetting plastics are molded. In an improved modification, called *hot runner molding*, the plastic is maintained in a fluid condition through large diameter runners all the way to the gate opening. Thus, scrap is essentially eliminated.

5.5 POWDER MOLDING

The art of dipping a preheated metal component in pulverized polyethylene has been practiced since 1940. It was patented by Genner under the terminology of fluidized bed coating in 1953. Nylon powder was one of the first plastics to be used but the materials have been extended to cellulosics, poly(vinyl chloride), acrylics, poly(chlorinated ethers), and epoxy resins. A similar process is used in the rotational molding of finely-divided polyethylene.

The metal part is usually coated in a two component tank in which the powdered resin held in place by a porous plate is mixed with a highly-compressed gas which is forced through the porous metal plate. The gas-powder mixture, which resembles a fluid, adheres to any metal part, which, after being preheated above the melting point of the resin, is placed in the fluidized bed atmosphere. A more uniform thickness results when the heated part is rotated. The final fusing of the polymer takes place in an oven or in a bath of molten salt.

In some cases, the coated part itself is used. In other cases, hollow articles, such as hobby horses and squeeze toys, are obtained by stripping the plastic from the metal form. Over two million pounds of powdered plastic were used for molding hollow parts in 1973.

5.6 CALENDERING

A calender consists of two or more hollow metal rolls rotating usually at equal speeds in opposite directions. This process, adapted from the rubber industry, continuously produces flat sheets of uniform thickness. The principal plastic used for calendering is poly(vinyl chloride), but acrylonitrile-butadiene-styrene copolymer (ABS) is also used to a limited extent. It is customary to use a Banbury heavy-duty mixer to supply a prefluxed mass of compounded plastic to the nip of the first two rolls.

The calendering process is comparable to that of extrusion in which the driven roll gap serves as the die lips and the coefficient of friction between the roll surface and the plastic stock serves as a force to move the plastic forward through subsequent rolls. It is customary to use four rolls that provide three nips which function successively for feeding, metering, and sheet formation. The thickness, speed, and temperature may be controlled at each nip. The output may vary from 600–10,000 lbs/hr depending on the size of the calender.

Embossed sheets may be produced by including an embossing roll as the last roll in the calender assembly. Laminates of fabric and plastic may be produced by a slight modification of the calendering technique. Over 100 million pounds of PVC sheet and film are produced annually by the calendering technique.

5.7 THERMOFORMING

Filled and unfilled polystyrene, cellulose acetate, poly(methyl methacrylate), styrene-acrylonitrile copolymer, and polypropylene may be extruded or calendered as sheet, heat softened, and formed into intricate shapes. In the simplest process, the sheet is softened at about 400°F and then drawn rapidly by a vacuum over a male form and allowed to cool. It is customary to stop the draw for a few seconds at the bottom dead center in order to assure a more uniform thickness. In one modification a plug is forced down on the heated sheet as it is pulled against the male form. A merry-go-round type assembly is used to automate this process.

In another modification, called *drape forming*, the male form is first forced upward against the warmed sheet prior to the application of a vacuum. In still another modification, compressed air is used to facilitate the *plug-assisted* vacuum forming process. In an adaptation of this thermoforming process, called *bubble pack*, a hot transparent film is pulled by a vacuum over a small object on a display card.

The sheet may be supplied from rolls or as twin rolls of different plastics in an automated version of this inexpensive process. For example, cigarette packs are produced at a rate of 80 halves a minute. These are then welded by heating the edges by friction to form 40 packs. The molds may be constructed from inexpensive materials such as wood or reinforced plastics. Large parts such as boats, housings and snowmobile bodies are readily produced by the thermoforming process.

BEHAVIORAL OBJECTIVES

1. Thermosetting plastics are usually molded by the compression molding process.

2. Labor costs of compression molding may be reduced by use of transfer

molding in which a hot preformed tablet is placed in a pot and the molten material is forced through an orifice into a hot mold cavity.

3. Compounded plastics may be converted to continuous forms such as pipe or rod by the extrusion process.

4. Bottles are produced by heating a hollow plastic section, closing one end of this parison, and using air to expand this molten plastic within a mold.

5. Thermoplastics are molded on a large scale using a highly automated injection molding process in which molten plastic is forced by a reciprocating plunger in a cooled, closed, two-piece mold. The thermoplastic scrap produced in this process may be recycled.

6. Metal parts may be coated by dipping the preheated parts in an aerated bed of plastic powder and heating the coated part to fuse the plastic and to assure a smooth surface.

7. Plastic film and sheet may be produced by passing a prefluxed mass of compounded plastic through a series of heated hollow rolls in equipment called a calender.

8. The thermoform process may be used to produce complex shaped structures by heat-forming plastic sheet of a male mold.

GLOSSARY

ABS: acrylonitrile-butadiene-styrene copolymer.
CAB: cellulose acetate butyrate.
Calender: a piece of equipment consisting of a series of hollow rolls used to produce film and sheet.
Draw: cavity depth in a mold.
DWV: drain, waste, and vent pipe.
Embossed: a raised textured surface.
Extrudate: plastic emerging from the die in an extruder.
Flash: molded resin that exudes outside the mold cavity.
Flight depth: the difference between the outside diameter and that of the cylinder root of an extruder screw.
Fluidized bed: an aerated plastic powder which resembles a fluid.
Gate: the small sized section of the runner at the entrance of the mold cavity.
GP: general purpose.
Land: a projection which controls the extent of travel of the plunger in a compression mold.
L/D: length to diameter ratio of an extruder.
Lead: the distance between ridges on an extruder screw.
Parison: a hollow specimen which is heated and expanded within a mold.
Pitch: the helical angle of an extruder screw.

Plasticate: to soften a plastic by mechanical means.

PSI: pounds per square inch of pressure.

Runner: the channel leading from the feed orifice to the mold cavity. The term also applies to the residual plastic in this channel.

Sprue: the stalk of cured thermoset plastic that remains in the feed orifice in transfer molding. The same term applies to thermoplastics in injection molding.

Toggle: a device in which pressure is applied on a knee joint.

Transfer molding: an automated type of compression molding in which a preform of plastic molding compound is forced from a pot into the hot molding cavity.

SUGGESTED QUESTIONS

1. How much pressure is required to compression-mold a part that has a thickness of 2 inches?
2. Which process is faster—ordinary compression molding or transfer molding.
3. What is the purpose of a breaker plate in an extruder?
4. What is the shape of the plastic material used to produce a bottle?
5. Which molding process is faster—compression or injection molding?
6. Why is the metal piece preheated before being immersed in a fluidized bed of aerated plastic powder?
7. Which equipment is more flexible for the production of plastic film or sheet—an extruder or a calender?
8. How would you produce a tub-like article with minimum equipment expense?

ANSWERS

1. 3,700 psi.
2. Transfer molding.
3. The breaker plate supports the screen pack and helps to assure intensive mixing of the plastic before it leaves the extruder die.
4. The parison is similar to a short section of pipe.
5. Injection molding.
6. To soften the plastic so that it will adhere to the metal surface.
7. A calender, since it can be used to produce coated fabric and embossed film.
8. By draping a warm sheet of a thermoplastic over a suitable inexpensive male mold, i.e., thermoforming.

chapter 6

Testing and Characterization of Plastics

Many applications of plastics are the result of actual trial and error tests. Plastics which do not meet these practical tests are rejected and those that meet them successfully are used for specific applications. However, when one knows the end-use requirements and is acquainted with the properties of various plastics, it is usually possible to eliminate much of the trial and error type of testing and to select appropriate materials for the new application.

Fortunately, standard tests are available for determining many properties of plastics. The use of standard tests is essential for quality control, for following the effect of actual use on the properties of plastics, and for the evaluation of modifications in formulations. Of course, it is always advantageous to develop and use additional tests which simulate the end-use applications.

6.1 TEST SPECIMENS

There are many cases in which the actual finished part is used in simulated tests as the test specimen. However, standard tests, such as those established by the American Society for Testing Materials (ASTM), usually require a standard specimen related to the finished part. Since many plastic test specimens are based on adaptations of those used in the metal, ceramic, or rubber industries, there may be little obvious relation between the test specimen and the part used in a specific application. Nevertheless, it is important that when standard tests are used, the test specimen be fabricated exactly in accordance with test specifications. It is customary to prepare a specified number of specimens in order to produce more reliable test data and to apply a statistical approach to the test data.

Many tests use *dog bone* or *dumbbell* test specimens. Since these are usually prepared by cutting the test piece from a larger piece, it is important to ascertain that all nonspecified notches are eliminated. ASTM test D647 describes molds for making bar and disc specimens. If the actual test piece is fabricated by thermal methods such as molding, extrusion, or thermoforming, it is advantageous to use a similar heat cycle in the preparation of the test sample and to minimize variations in heat treatment among the individual test specimens. Regardless of any lack of relationship between the shape of the test specimen and the finished part, it is important that all test specimens be representative.

6.2 CONDITIONING OF TEST SPECIMEN

The service conditions may vary from temperatures of $-40°$ to $60°C$ and from a relative humidity of from 0 to 100% but the tests are usually performed at specified conditions of temperature and percent humidity. ASTM test D618 specifies that the test specimens be stored and tested at $23°C$ and 50% humidity, and these conditions are considered as *standard test conditions* unless others are specified. The tolerances are usually $\pm 2°C$ and $\pm 5\%$ humidity.

6.3 ENVIRONMENTAL TESTS

Plastics, as are other materials of construction, are exposed to various environments during use. The least severe exposure is to the atmosphere. Most plastics are resistant to mildly-corrosive atmospheres, such as aggressive sea coast conditions. However, many plastics with weak chemical bonds, such as those of chlorine to carbon in poly(vinyl chloride) (PVC), are adversely affected by ultraviolet radiation in the atmosphere. When used as containers or immersed in liquids, plastics are exposed to many different solvents and corrosives, and it is important to test the effect of these liquids on specific plastics before use.

6.3.1 Environmental Stress Cracking

Some polymers, such as low density polyethylene (LDPE), crack when exposed to polar solvents or aqueous detergents. Since polyethylene bottles are often used as containers for such liquids, it is important that the effect of the contents on the plastic be determined. *Environmental stress cracking* (ESCR) tests have been developed for sheet, pipe, and blow-molded containers. The ASTM test D-1693 was developed by the Bell Telephone Laboratories.

In this specific test, recommended for LDPE only, specimens measuring 1 in $\times \frac{1}{2}$ in $\times \frac{1}{8}$ in are annealed in boiling water for 1 hr and then conditioned at room temperature. The specimens are notched with a razor blade and bent in a U shape at 80° within a brass channel and the assembly is immersed in a test tube. The immersed specimens are inspected periodically and the time for rupture is recorded. After completion of the test, the percent of failures is recorded.

6.3.2 Weathering Tests

In this test, ASTM test D-1435, plastic panels are mounted at a 45° angle on outdoor racks facing south and the exposed samples are compared with unexposed samples to determine the effect of exposure.

These time-consuming weathering tests may be accelerated by using cyclic exposure conditions which attempt to simulate natural weathering conditions. Devices such as the weatherometer may be used to hasten weather-like deterioration of plastics. Some correlation may be demonstrated for test specimens exposed to accelerated tests. However, weather itself is variable seasonally and geographically and hence little correlation exists between the natural and accelerated weathering processes. The accelerated weathering test is described by ASTM test G-23.

6.3.3 Solvent Effects

When plastics are used as containers, gears, clothing, or furniture, they must be resistant to specific solvent environments. Some information on solvent resistance may be obtained by reference to the solubility parameters of the plastic and liquid. When these are dissimilar, such as in the case of silicones and water, the plastic is not attacked by the solvent. Objective data may be obtained by actual immersion of plastic specimens in specific solvents. Excessive attack usually can be observed by noting changes in the surface, shape, or swelling. More subtle changes may be noted by the rate of change in weight, hardness, or flexural strength of exposed samples.

Plastics, such as polyethylene, that are characterized by relatively-low solubility parameter values ($\delta = 8.0H$) are essentially unaffected by water. However, many plastics absorb small amounts of water. The effect may be noted by observing changes in dimensions or in electrical properties. In the standard ASTM test D-570, standard-sized test specimens or coupons are immersed in water and changes in the weight of cloth-wiped samples are noted.

6.3.4 Oxidation

The oxygen in air will oxidize the polypropylene, polystyrene, and low-density polyethylene, particularly in the presence of ultraviolet light or moderate heat. Excessive oxidation of colorless plastics will cause a yellow discoloration. The effect of less drastic oxidation may be observed by noting changes in the power factor of these materials.

Many plastics may be attacked by oxidizing acids, such as nitric acid, chromic acid, or concentrated sulfuric acid. Effects such as air oxidation may be noted by visual observation or by measuring the change in the power factor.

6.3.5 Corrosion

The effect of corrosive liquids on plastics may be determined by observing the change in weight of test specimens immersed in the corrosive liquid in accordance with ASTM test D-543. The rate of change of hardness, flexural strength, and electrical properties may also be used to follow degradation of plastics in a corrosive environment.

6.4 DIMENSIONAL TESTS

The *density* or *specific gravity* of a plastic specimen may be accurately and rapidly determined by placing it in a density gradient column at 23°C and noting its level at equilibrium. Columns in which the density of the liquid decreases with the height are described by ASTM test method D-1505.

The plastic specimen is attached to a fine wire and weighed in air and submerged in water, in ASTM test procedure D-792, which is based on Archimedes' principle. The specific gravity of a plastic molding compound may also be calculated by weighing a measured volume of powdered plastic in a pycnometer. The specific gravity data for granular molding powders are reported as bulk density or *apparent density*.

6.5 THERMAL TESTS

The heat-softening temperature may be determined by the Vicat test which is described by ASTM test D-1525. In this penetration-type test, the temperature of the specimen is increased at a rate of 50°C per hour, and the temperature is recorded when a standard needle with a surface area of 1 mm^2 under a load of 1 kg penetrates the specimen to a depth of 1 mm.

The coefficient of cubical thermal expansion is determined by noting the rise of mercury in a capillary tube when a standard-sized plastic rod is heated in a *dilatometer*. In this test, described in ASTM procedure D-864, corrections are made for the thermal expansion of glass and mercury, and the coefficient of expansion is calculated from the slope of the linear line when expansion is plotted against the temperature. Changes in the slope may be observed at transition points such as the glass-transition temperature.

ASTM test D-696 describes a measurement of the linear expansion of a standard bar. The change in length as noted on a dial gauge is used to calculate the linear coefficient of expansion. Sophisticated instruments are also available for this measurement.

The effect of low temperatures on the brittleness of a plastic specimen is determined by lowering the temperature of a cantilevered specimen and observing the effect of impact. This test is described in ASTM procedure D-746. Cold-bend and cold-flex temperatures are determined by winding a test piece on a mandrel and by subjecting a specimen to torque tests at low temperatures.

There are many tests for measuring the *flammability* of plastics. In ASTM D-635, a test specimen is heated for 30 sec with a Bunsen burner and any residual burning after removal of the flame source is noted. The lowest ambient temperature at which ignition of a plastic specimen takes place is noted in ASTM test D-1929. The ultimate test is a tunnel test such as the Underwriters Laboratories Standard 723 test. The relative resistance of plastics to flame may also be estimated by determining the minimum of oxygen concentration required for ignition.

One of the most widely used thermal tests is the ASTM D-648 *deflection temperature test*. The application is shown in Fig. 6-1. In this procedure, a standard specimen mounted as a simple beam and a load of 66 or 264 psi is placed at the center. The bar is heated at a rate of 2°C/min, and the temperature at which the bar deflects 0.010 in is recorded as the deflection temperature.

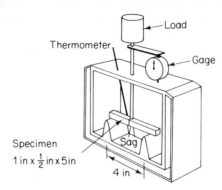

Fig. 6-1 Apparatus for measuring the deflection temperature.

6.6 MECHANICAL TESTS

Many standard mechanical tests have been devised in an attempt to predict the performance of specific plastics. The ratio of stress to strain, which is a measure of stiffness, can be used to interpret many of the physical properties of plastics. Therefore, a discussion of these relationships is in order.

6.6.1 Stress Strain Curves

According to Hooke's law, the tensile stress is proportional to the strain and the proportionality constant is the modulus of elasticity. Plastics which are viscoelastic materials are not Hookean but much information can be gained from a study of their stress-strain curves. Carswell and Nason used the curves shown in Fig. 6-2 to classify plastic materials as soft and weak, hard and brittle, soft and tough, hard and strong, and hard and tough. A generalized-tensile stress-strain curve is depicted in Fig. 6-3.

An examination of the curve for a hard, tough plastic, such as plasticized cellulose acetate, will show that this plastic has a high modulus up to its yield point. It continues to elongate after passing the yield point until it fails at the ultimate stress. The plastic exhibits hookean behavior up to its yield point and its elongation in this elastic range is recoverable.

In contrast, the elongation observed after the yield point is not completely recoverable. The area under the curve in the plastic range between the yield point and the ultimate strength is a measure of toughness. It is important to note that all stress-strain curves are time and temperature dependent. Thus, one must use standard testing rates and temperature for all mechanical tests. A slower rate of stress will yield a lower slope and therefore a lower modulus of elasticity. The opposite is also true. A comparable decrease in the modulus of elasticity will be noted when the temperature is increased.

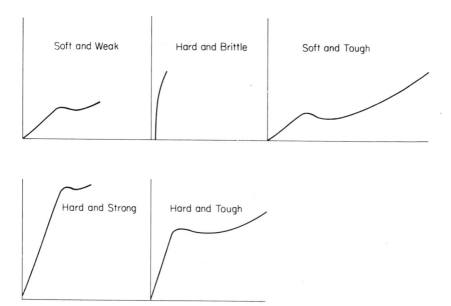

Fig. 6-2 Tensile stress-strain curves for several types of polymeric materials.

Actually the curve for soft and tough plastics is a typical stress-strain curve for material such as plasticized PVC. These curves will not be obtained if the temperature is lowered excessively. A curve such as the hard and brittle curve, which is typical for polystyrene, could be obtained when plasticized PVC is tested at very low temperatures. Soft plastics, such as polyisobutylene,

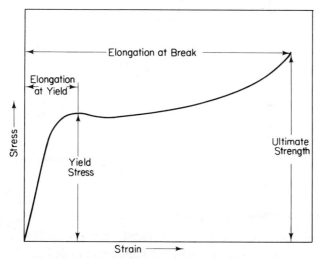

Fig. 6-3 Generalized stress-strain curve.

have stress-strain relationships similar to the soft and weak materials. Strong rigid plastics, such as unplasticized PVC, have properties similar to those described as hard and strong.

Some information on the mechanical properties of polymers may also be obtained from the relative decrease in cross-section when a plastic specimen is stretched. The ratio of this contraction to elongation, called *Poisson's ratio*, is about 0.5 for elastic materials and decreases to about 0.25 for very stiff materials.

6.6.2 Tensile Strength

The tensile test specimen is usually a $\frac{1}{8}$ in. thick dogbone piece with a central cross-section of $\frac{1}{2}$ in, as described in ASTM procedure D-638. Figure 6-4 shows that the specimen is clamped in the testing machine jaws and the lower movable jaw is moved downward at a fixed rate. The stress or force that resists the elongation process is plotted against the elongation or strain. As previously described, this ratio is the modulus of elasticity or tensile modulus. The tensile test is the most widely-used test for evaluating the strength of plastics. Plastics parts must be designed to accommodate stress considerably below that of the yield point.

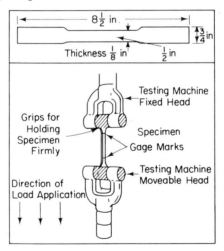

Fig. 6-4 Sketch of tensile test specimen and testing machine jaws holding specimen.

6.6.3 Flexural Strength

In accordance with ASTM procedure D-790, flexural testing consists of measuring the force required to cause a simple beam to deflect 5%. It is not possible to calculate the flexural or transverse strength of flexible plastics since they usually do not fail under a deflection of less than 5%. Therefore, the flexural strength reported for many thermoplastic materials is the force required to cause a deflection of 5%.

6.6.4 Impact Strength

Some information on the toughness of a plastic specimen may be obtained from the area under the stress-strain curve. However, data on toughness are usually supplied from *Izod* or *Charpy impact tests*, which, though controversial, will continue to be used until a more universally-acceptable alternative test is available. In the Izod test, described in ASTM procedure D-256, a swinging pendulum is used to strike a cantilevered notched specimen, as illustrated. (See Fig. 6-5.) The impact strength is reported as ft/lb/in of notch for breaking a 1 in. specimen.

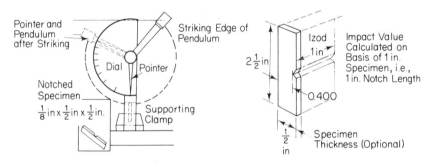

Fig. 6-5 Izod test.

In an alternative test, described in ASTM procedure D-1822, an unnotched cantilevered specimen is struck by a swinging pendulum. In the *Charpy impact test*, the notched specimen is supported as a simple beam by two anvils and the swinging pendulum strikes behind the notch on the unnotched side of the test specimen. It is important to note that tough notch-sensitive plastics such as nylon and acetals are characterized by relatively low-impact strengths based on these standard tests.

In another impact test, described in ASTM test D-1822, a $2\frac{1}{2}$ in long bar is mounted between a pendulum head and a cross-head clamp on the pendulum so that the specimen swings past a fixed anvil after it is released. The clamp is halted on contact with the anvil but the pendulum head continues forward and the energy loss is recorded.

6.7 HARDNESS

Hardness is a measure of resistance to penetration, abrasion, and scratching. While the standard hardness tests are primarily a measure of resistance of a plastic to penetration, *scratch resistance* is often a more significant property of plastics. In accordance with ASTM test procedure D-618, indentation hardness may be determined by measuring the depth of penetration of a loaded indenter. The instrument used is called a *Shore Durometer*, and the hardness is reported in dimensionless units on a scale of 0 to 100. Hardness may also be reported as Rockwell hardness. In this test, described in ASTM procedure D-785, the indentation is measured after the removal of a loaded ball.

Abrasion resistance may be determined by measuring the rate of weight loss caused by the mechanical rubbing of the surface of a plastic. The relative resistance to scratching may be determined by noting the scratching effect of graded minerals using Mohs' hardness scale in which the following values are assigned to the standards listed: 1. talc, 2. gypsum, 3. calcite, 4. fluorite, 5. apatite, 6. orthoclase, 7. quartz, 8. topaz, 9. corundum, 10. diamond.

6.8 ELECTRICAL TESTS

6.8.1 Arc Resistance

Arc resistance or resistance to tracking is defined as the ability of a plastic to withstand a high voltage-low current discharge across its surface. In ASTM procedure D-495, the arc resistance is reported as the interval between the formation and disappearance of flashes from the arc of the high voltage discharge.

6.8.2 Dielectric Constant and Dissipation Factor

Dielectric constant or permittivity is determined by ASTM procedure D-150. This value is the ratio of the capacitance of a condenser made from a plastic material to that of a similar condenser using air as the dielectric. In this test, the electrodes are placed on opposite sides of a plastic sheet and the capacitance is measured by use of an electric bridge circuit.

The dissipation factor is a measure of the conversion of reactive power to real power or it may be defined as the ratio of the in-phase power to the out-of-phase power.

6.8.3 Dielectric Strength

This test, described in ASTM procedure D-149, is equal to the maximum voltage which a 1 in thick sheet of plastic can withstand without failure for 1 min. The value is determined by placing a thin sheet of plastic between electrodes and noting the minimum voltage required for breakdown as evidenced by the passage of current through the sheet. The dielectric strength or dielectric breakdown voltage is calculated by dividing the maximum voltage withstood for 1 min by the thickness of the sheet in mils.

6.8.4 Power Factor

When a plastic sheet is used as a dielectric in alternating current (AC), the ratio of the energy loss to the total energy passing through the capacitor is called the *power factor*. This test is described in ASTM procedure D-150. Since considerable heat is developed when plastics with high power-factor values are used as capacitors, the heat generated may be used to soften or melt thermoplastics such as PVC. In contrast, plastics such as polyethylene with low power factor values are suitable for use as electrical insulators. The loss factor is obtained by multiplying the power factor by the dielectric constant.

6.8.5 Volume Resistivity

Volume resistivity is the reciprocal of the ratio of conductivity or potential gradient parallel to the current flowing through the plastic. In ASTM procedure D-257, a potential gradient is applied to two electrodes separated by the plastic specimen. The volume resistivity or resistance to leakage of electrical current is the product of length of the conductor times its resistance divided by its cross-sectional area. When the length and area are unity, the volume resistivity is the electrical resistance between opposite faces of a 1 in cube.

6.9 OPTICAL TESTS

6.9.1 Index of Refraction

By definition, the index of refraction or the extent of bending of light in a plastic is the ratio of the velocity of light in a vacuum or usually in air to its velocity in the plastic material. This characteristic value of transparent plastics may be determined by ASTM procedure D-542 using an Abbé refractometer. In this test, a polished plastic specimen in the refractometer is wetted with a drop of a selected liquid and the index of refraction is determined in the usual manner. The liquid must not soften the plastic and should have an index of refraction slightly higher than that of the plastic. Alpha-bromonaphthalene is used as the contacting liquid for PVC and phenolic resins.

6.9.2 Haze and Luminous Transmittance

Haze or milkiness of a plastic may be measured by determining the amount of light deviating by more than 2.5° from the direction of the transmitted light. In ASTM procedure D-1003, the amount of light passing through a film to a reflecting surface of a sphere is measured by use of a photoelectric cell. This measurement is then repeated in the hazemeter where the beam of light falls on an absorptive dark mat and the scattered light is measured. The percent haze is calculated by multiplying the ratio of the first and second transmittance by 100. Luminous transmittance is defined as the ratio of transmitted to incident light.

6.9.3 Luminous Reflectance

In ASTM test D-791, a Hardy-type spectrophotometer is used to determine the luminous reflectance, transmittance, and color properties of nonfluorescent plastics. The reflectance of the plastic is compared with that of white chalk. Gloss is also a measure of the ability of a plastic to reflect light.

6.10 NONDESTRUCTIVE TESTING

While the tests described previously require the testing and usually the destruction of the specimen, there is a need to preserve plastic parts that have met test requirements and to withdraw from service those that fail to meet the minimum specifications. These objectives are met by nondestructive tests (NDT), which are designed to obtain data without impairing the usefulness of the plastic part. Such tests have been developed to a high degree of reliability for reinforced plastics.

The most important and least sophisticated NDT is visual inspection where the observer notes any obvious defects. This is often aided by use of both reflected and transmitted light. Ultrasonic tests with frequencies in the range of 100 khz to 25 mhz are used for the detection of porosity, voids, or nonuniform areas. The manual coin-tapping technique used by the artisan has been made more sophisticated, more reproducible, and more reliable by utilizing electromagnetic techniques to form eddy currents so that the acoustical responses may be monitored.

Metallic contaminants may be located by conventional x-rays, beta radiation, or thermal neutrons. Some of the radiographic techniques have been monitored by fluoroscopy or television. Since exposure of personnel to these techniques is hazardous, microwave or radar techniques are often preferred for NDT.

Other NDT techniques include electrical, thermal, holographic, and nuclear quadrupole resonance (NQR). The principal electrical tests are based on the observation of any change in dielectric constant and dissipation factor with time. The principal thermal test consists of heating the plastic object to a uniform temperature and noting any nonuniform cooling of the surface.

The holographic NDT test consists of monitoring holographic interference patterns on the surface of the plastic object while it is subjected to thermal or mechanical stress. In order to use NQR techniques, it is necessary to incorporate a compound such as copper(I) oxide which has a nuclear quadrupole moment. Since any applied force on the plastic part will then cause a shift in the measured resonance frequency, flaws and nonuniform areas in the plastic may be detected.

6.11 CHARACTERIZATION

6.11.1 Molecular Weight Determination

Some naturally occurring polymers, such as proteins, and some synthetic polymers, such as some prepared by anionic polymerization techniques, consist of products with one specific molecular weight and are called *monodisperse*. In contrast, naturally occurring polymers, such as cellulose, and most synthetic polymers consist of a mixture of products of different molecular weights and are called polydisperse. The ratio of the weight average molecular weight and the number average molecular weight is a measure of polydispersity of a polymer. This value is equal to 1 for a monodisperse polymer and greater than 1 for polydisperse polymers. Techniques are available for determining both weight average and number average molecular weights.

6.11.2 Colligative Techniques

Properties of solutions which depend only on the number of solute particles present, such as vapor pressure, boiling point, freezing point and osmotic pressure, are independent of the composition of these particles and are called *colligative properties*. Since the vapor pressure of a solution is always less than that of the solvent and since the resultant vapor pressure lowering is related to the number of solute particles present, vapor pressure lowering can be used to measure the molecular weight of solutes. The boiling points and freezing points are also affected by the number of solute particles present. Thus the boiling point increase and the melting point decrease may also be used to measure the molecular weight of small molecules. The ebulliometric and cryoscopic techniques have also been used to determine the molecular weight of macromolecules, but extremely precise measuring techniques are required.

The mole fraction of the solute of a 1% solution of a polymer such as polystyrene in benzene is 7.8×10^{-5}. The presence of this number of particles of solute will depress the vapor pressure by 7.3×10^{-3} torr, increase the boiling point by 2.5×10^{-3}°C and decrease the melting point by 5.1×10^{-3}°C. Such measurements require extremely sophisticated techniques. Fortunately, the magnitude of the measured values for the increase in osmotic pressure is much greater than for other colligative properties. Therefore, osmometry is the preferred technique for determining the number average molecular weight of polymers which do not diffuse through a semipermeable membrane.

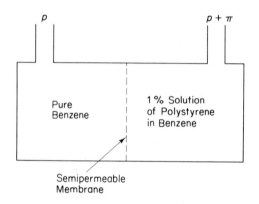

Fig. 6-6 Sketch of an osmometer.

The osmotic pressure exerted by a benzene solvent separated by a semipermeable membrane from a 1% solution of polystyrene in benzene is 2.5×10^4 dynes/cm². As shown in Fig. 6-6, this osmotic pressure (π) would cause the solution to rise and the solvent to drop in capillary tubes. The difference in the height of the liquid in the two capillary columns for this solution would be 28 cm.

The relation of the osmotic pressure (π) to the number average molecular weight ($\overline{M}_n$) is shown in the following equation in which C is the concentration, T the temperature in degrees kelvin (°K) and R the gas constant:

$$\pi = \frac{RTC}{\overline{M}_n}$$

Fig. 6-7 Relationship of osmotic pressure to molecular weight.

Freshly-extruded cellophane film that has been dehydrated by washing with ethanol may be used as a semipermeable membrane. The membrane is then clamped between two metal slabs in which interconnecting concentric rings have been machined on the internal faces. These cavities are connected to capillary tubes. The use of membrane osmometry is restricted to polydisperse systems in which $\overline{M}_n$ is greater than 50,000.

Since the properties of a solution of a polymer in a good solvent are not independent of concentration, it is necessary to determine the osmotic pressure for several different concentrations and then to extrapolate to zero concentration in order to eliminate the effect of the intermolecular forces between solvent and solute. Thus, when π/RTC is plotted against C, and extrapolated to 0, the intercept is $1/\overline{M}_n$, as shown in Fig. 6-8. The slope

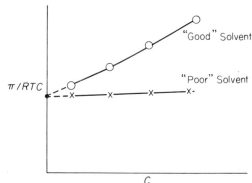

Fig. 6-8 Plot of π/RTC vs C used to determine $1/\overline{M}_n$ in osmometry.

of the curve is a measure of polymer-solvent interaction and is greater for a good solvent.

The molecular weight of lower molecular weight polymers may be determined by vapor phase osmometry, illustrated in Fig. 6-9. This technique is dependent on the precise measurement of the temperature difference between an evaporating drop of a solvent and a drop of the polymer solution. The instrument is calibrated by using solutions of known molecular weights.

6.11.3 Light Scattering Techniques

Because of the characteristic rapid motion of molecules in a liquid, a transitory change in the index of refraction occurs which is evidenced by light scattering. This effect is related to the Tyndall effect observed when a beam

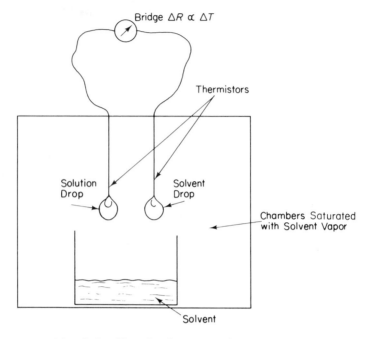

Fig. 6-9 Sketch of a vapor phase osmometer.

of light encounters dust particles in a dark room. It is enhanced when large molecules, such as polymers, are present in a solution.

Thus, the average weight of the solvent molecules may be determined by observing the intensity of the incident and scattered light. Accordingly, it is possible to calculate the end-to-end distance of solute molecules by measuring the ratios of the intensities of scattered light to the incident beam at two different angles, such as 45° and 135°. The value obtained by this technique is the weight average molecular weight $\overline{M}_w$.

6.11.4 Viscometry

When a macromolecule is dissolved in a good solvent, the polymer chain is extended to its full contour length and the viscosity of the solvent is increased. Since the polymer chains tend to coil in poor solvents, the addition of a macromolecule has less effect on the viscosity of the poor solvent. However, because of the molecular size and changes in dimension, the viscosity increase in both systems is many times greater than that observed when a comparable mass of a nonpolymeric solute is added.

The increase in viscosity of a polymeric solution may be used as a measure of the average molecular weight. However, unlike osmometry and light scattering techniques, viscometry does not provide absolute values for molecular weight. Viscosity is readily measured and viscometry may be used to determine molecular weights after a system has been calibrated using absolute data. Even in the absence of this calibration, the relative viscosities of solutions of specific polymer are valuable data for estimating relative molecular weights.

The viscosity (η) of a solvent and the polymer solution may be easily determined by measuring the efflux time for capillary flow in a viscometer tube. The ratio of the efflux time of capillary flow for the polymer solution to that of the solvent is equal to the ratio of their viscosities and this solution/solvent viscosity ratio is designated as the relative viscosity η_r or *viscosity ratio*.

The term $\eta_r - 1$, which varies with concentration, is called the *specific viscosity* (η_{sp}). The ratio of the specific viscosity to concentration (η_{sp}/C), which is less dependent on concentration (*C*), is designated as the *reduced viscosity* or preferably as the *viscosity number*.

The value obtained by the extrapolation of the viscosity number to zero concentration or limiting viscosity number is an intensive property from which molecular weight data may be determined. The limiting viscosity number is also called the *intrinsic viscosity* or *Staudinger index* and is designated as [η]. Another widely-used term is *inherent viscosity*, which is equal to 2.3 log η_r/C.

Intrinsic viscosity is a function of the average molecular weight ($\overline{M}$), as shown by the Mark-Houwink equation in Fig. 6-10. The values for the

$$[\eta] = K\overline{M}^a$$

Fig. 6-10   Mark-Houwink equation.

constants *K* and *a* at a specific temperature may be determined experimentally using solutions of polymers of known molecular weight. The constant *a* is related to the shape of the molecule in a specific solvent. Both values can be found in handbooks such as the Polymer Handbook, Interscience Publishers, John Wiley and Sons, N.Y. (1966).

6.11.5 Gel Permeation Chromatography

Gel permeation chromatography (GPC) or gel filtration is one of the more useful techniques for fractionation of polymers and for the determination of molecular weights. This apparatus consists of two parallel columns in which the stationary phase usually consists of swollen small gel particles of a crosslinked polystyrene. The mobile phase, which passes downward through one of these columns, is a solution of polymer such as a solution of polystyrene in tetrahydrofuran (THF). The pure solvent is eluted through the other column.

The flow rates through these columns are adjusted to about 1 ml/min and the difference in the index of refraction of the solution fraction and the solvent is recorded and related to the average molecular weight by calibration using samples of known molecular wieght.

Since the pore size of the solvent swollen gel can be as small as 10Å and as large as 10^7Å, a wide range of molecular weights may be separated by this GPC technique. The column efficiency or plate count per ft of column (P) is determined by the elution of a pure liquid such as o-dichlorobenzene. The efficiency is often reported as $1/P$, that is, the height equivalent in feet to one theoretical plate (HETP). The plate count is also directly proportional to the square of the eluted volume (Ve), as shown in Fig. 6-11.

$$P = \frac{1}{F}\left(\frac{4Ve}{d}\right)^2$$

Fig. 6-11 Equation for calculating the plate count in GPC.

This figure illustrates that the plate count (P) is also inversely proportional to the length of the column in feet (F) and to the square of the base line of the GPC curve in ml (d). The latter is the length of the base of the triangle obtained when one draws straight lines tangent to the curved sides of the GPC peak.

The hydrodynamic volume (V) is directly proportional to the weight average molecular weight ($\overline{M}_w$). By the equation in Fig. 6-12, one sees that the proportionality GPC constant for polystyrene in THF at 25°C is 1.5×10^{-2} ml/g.

$$V = 1.5 \times 10^{-2} \, \overline{M}_w^{1.70}$$

Fig. 6-12 Equation for calculating the molecular weight of polystyrene from GPC data.

Thus, a linear calibration curve may be obtained when the logarithm of the hydrodynamic volume (V) is plotted against the elution volume (Ve). Dextran gels have been used to fractionate water-soluble polymers. Crosslinked polymers are now used routinely in GPC for the determination of the molecular weight distribution of water insoluble polymers.

6.12 SPECTROSCOPY

In addition to techniques discussed previously under tests, the principal methods used for the characterization of plastics are infrared absorption spectroscopy, nuclear magnetic resonance spectroscopy, and thermal analysis.

6.12.1 Infrared Absorption Spectroscopy

The first instrument for determining and recording the residual spectra resulting from the absorption of energy by organic molecules in the infrared region (IR) was marketed in 1945. This technique is particularly useful for the characterization of organic polymers that may be tested as film or powder. A modification, called *attenuated total reflectance spectroscopy* (ATR), may be used for the characterization of thick samples. The latter technique may also be used for nondestructive testing (NDT).

The infrared region of the spectrum is defined as that having wavelengths ranging from 1–500 μ. The micron (μ) is equal to 10^{-6} m and the term micrometer is preferred. Unfortunately, in addition to the use of other linear units such as the angstrom (Å) (1Å = 10^{-8} cm), this range is also reported in terms of wavenumbers ($\bar{v}$) or kaysers which have the unit of reciprocal centimeters (cm^{-1}). The relationship between reciprocal centimeters and wavelengths (λ) is the velocity of light (C) divided by the frequency (v) of the radiation. Thus, the wavenumber ($\bar{v}$) is equal to the reciprocal of the wavelength (λ^{-1}), i.e. $1/\lambda = v/C = \bar{v}$. Accordingly, the infrared range is also 10,000–20 cm^{-1} or kaysers.

The absorption of infrared radiation or any other electromagnetic radiation is quantized, i.e. the energy absorbed as a result of transitions between quantum levels in the sample is equal to discrete values. Thus, the change in energy (ΔE) between these levels is proportional to the frequency (v) and the proportionality constant is Planck's constant (h) (6.63 × 10^{-27} erg sec), i.e. $\Delta E = hv$.

These low-energy infrared waves may cause characteristic changes in the vibrational and rotational states in organic molecules. While relatively simple spectra are observed for simple molecules, polymers, which are polyatomic molecules, produce complex spectra that may be difficult to interpret. Therefore, the styrene segment ($C_6H_5(CHCH_2)$) with 16 atoms, present in the polystyrene molecule, has 42 degrees of vibrational freedom. Its infrared spectrum is used as a standard for the calibration of infrared spectrophotometers. One can account for the vibrations responsible for almost all of the bands in this standard spectrogram.

Since groups in molecules have characteristic infrared spectra, this type of spectroscopy may be used to detect specific groups in polymers. The intensity of these absorption bands is proportional to the concentration of these typical groups. Thus, infrared spectroscopy may be used both as a versatile qualitative and quantitative analytical tool.

6.12.2 Nuclear Magnetic Resonance Spectroscopy

In contrast to the high energy radiation associated with x-rays, ultraviolet light, and, to a lesser extent, with visible light, the long radio frequency waves (RF) which cause changes in nuclear magnetic resonance spectroscopy (NMR) have very low energy. However, this energy, though less than that in the infrared region, is sufficient to orient the spin of the nucleus of the hydrogen or fluorine atoms. Thus, one may use NMR for the analysis of most organic polymers.

The principles of NMR have been known for over 40 years and the first experimental techniques for NMR spectroscopy were demonstrated shortly after World War II. Instruments with different degrees of sophistication are now available commercially. This technique may be used with polymeric solutions to demonstrate the difference in structure of polymeric hydrocarbons such as LDPE, HDPE, and PP.

NMR spectroscopy has been used to identify isotactic, syndiotactic *PP*, and heterotactic or atactic sequences, and the cis and trans content of polymers. Since the flipping of the nucleus in a magnetic field is slower in crystalline, reinforced and crosslinked polymers, NMR wide-line studies have been used to show the degree of crystallinity, the effectiveness of fillers, and the extent of crosslinking in polymers. Since the carbon-13 isotope is present to the extent of 1.1% in all organic polymers, carbon-13 NMR is also used as a sophisticated characterization technique.

6.13 THERMAL ANALYSIS

6.13.1 Pyrolysis Gas Chromatography

Since many polymers decompose when heated, an analysis of the characteristic pyrolytic products may be used for their characterization. As gas chromatographs are readily available, the off gases are usually analyzed by gas chromatography (GC). It is customary to pyrolyze a thin polymeric film for a few seconds on an electrically heated platinum coil and to pass the gaseous products along with a carrier gas such as helium through a GC column with an appropriate packing.

The gaseous components with the least affinity for the packing pass through the column first. These and those with longer retention times may be detected with appropriate devices as they leave the column. The peaks on the recorded charts (pyrograms) characteristic of the decomposition products may be identified by comparison with the peaks having similar retention times from known compounds. This technique may be used for polymer identification and for the quantitative analysis of components of polymers.

6.13.2 Differential Thermal Analysis

The principles of differential thermal analysis (DTA), sometimes called *thermal spectroscopy*, were demonstrated by Le Chatelier in the nineteenth century. Many different types of DTA instruments are now available commercially. This technique, which measures the change in thermal energy of a substance as a function of temperature, has been widely used for the characterization and investigation of polymers.

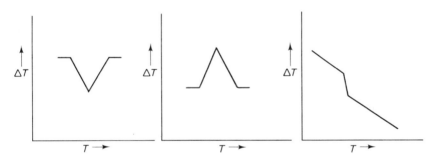

Fig. 6-13 Simulated thermograms showing endothermic, exothermic, and thermal transition changes respectively.

As shown in Fig. 6-13, characteristic thermograms are obtained when the difference in temperature (ΔT) is recorded when a polymer sample and an inert control are heated at a programmed rate in an inert atmosphere. A V-shaped band is noted when heat is absorbed, and an inverted V is observed when heat is released.

Since characteristic thermograms are obtained for polymers in a mixture, DTA may be used for qualitative and quantitative analysis of mixtures of polymers. It is of interest to note that the thermograms of mixtures of polymers differ from those of related copolymers. DTA may also be used to differentiate between random, block, and graft copolymers.

In a modification of this technique, called *thermal gravimetric analysis* (TGA), the loss in weight with the increase of temperature and/or time is recorded. TGA may be used to characterize polymers and to evaluate their stability at higher temperatures.

BEHAVIORAL OBJECTIVES

In the absence of appropriate testing and characterization, plastics may fail under actual use conditions. Hence, it is important that adequate testing and characterization programs be established by producers and fabricators of plastics to minimize the occurrence of preventable failures and to assure public acceptance of plastics.

After reading this chapter, you should understand the following:

1. Testing of plastics is usually done on conditioned test specimens using standard test procedures.
2. Exposure tests for plastics include exposure to the environment which may be sunlight, detergents, solvents, oxidizing agents, and corrosives.
3. Thermal tests on plastics include penetration tests, expansion tests, and high temperature heat deflection tests.
4. The area under a stress-strain curve is a measure of toughness.
5. Typical stress-strain curves for plastics are available and these may be used for comparative purposes.
6. The tensile modulus is the ratio of the elongation to tensile stress.
7. The flexural strength which is the resistance of a test specimen to bending forces is an important property for rigid structural plastics.
8. Plastics, like dishware or tools that may be dropped, must have good resistance to impact.
9. Since plastics are good electrical insulators, electrical tests are extremely important.
10. Since the plastic test specimen is still available for use after testing, nondestructive testing procedures are highly desirable.
11. Considerable information on polymer structure and molecular weight range may be obtained by the use of more than one test for molecular weight.
12. The number average molecular weight is obtained from osmotic weight data by plotting at several different concentrations and extrapolating to zero concentration.
13. The weight average molecular weight is obtained from light scattering data by plotting at several different concentrations and extrapolating to zero concentration.
14. Since the viscosity of a solution is related to the molecular weight of

the polymer solute, the molecular weight may be calculated providing values are available for the constants in the Mark-Houwink equation.

15. Molecules of all sizes including macromolecules may be separated according to size by use of gel permeation chromatography.

16. Infrared spectroscopy may be used to characterize macromolecules both qualitatively and quantitatively.

17. NMR spectroscopy may be used to determine the tacticity of macromolecules.

18. The composition of copolymers can be determined routinely by pyrolysis G.C.

19. DTA may be used to determine Tg and Tm for polymers. Mixtures of polymers and block copolymers will exhibit independent Tg values.

GLOSSARY

ASTM: American Society for Testing Materials.
Bulk density: density of molding powders or apparent density.
Bulk factor: density of molded specimen or apparent density.
Cantilever: a beam supported at one end only.
Charpy: an impact test.
Colligative properties: those dependent on the number rather than kind of molecules in a solution.
DTA: differential thermal analysis.
Elastic limit: the point on the stress-strain curve below which stressed material can recover its original dimensions when the stress is removed, i.e. the yield point.
$[\eta]$ = limiting viscosity number or intrinsic viscosity.
η_r = relative viscosity.
η_{red} = reduced viscosity.
η_{sp} = specific viscosity.
Flexural strength: the resistance of a test specimen to bending forces.
G.C.: gas chromatography.
G.P.C.: gel permeation chromatography.
Hookean: obeys Hooke's law.
Impact strength: the resistance of a test specimen to a sharp blow.
IR: infrared.
Izod: an impact test.
Mark-Houwink equation: $[\eta] = K\overline{M}^a$
NDT: nondestructive tests.
NMR: nuclear magnetic resonance.
NQR: nuclear quadrupole resonance.
$\bar{\nu}$: wave number.

π: osmotic
R: ideal gas constant = 2 cal per mole.
Strain: the extent of deformation of a stressed test specimen.
Tensile strength: the resistance of a test specimen to elongation forces.
TGA: thermogravimetric analysis.
Yield point: the highest load to which a plastic can be subjected without failure.

SUGGESTED QUESTIONS

1. If you needed guidance on plastic testing procedures, where would you seek such information?
2. Why is the measure of the coefficient of expansion of a plastic important?
3. Which plastic should be selected for high temperature applications, (a) one with a heat deflection temperature below 150°F or (b) one with a heat deflection temperature above 300°F?
4. Higher moduli values can be obtained by using faster mechanical test cycles. Is this technique sound?
5. Which plastic would be tougher, (a) one having a small area under the stress-strain curve, or (b) one with a relatively large area?
6. Why is high flexural strength important in structural plastics?
7. What is the failure that occurs when a dropped plastic shatters?
8. What hardness test is most significant for plastic glazing?
9. Name a few household products or appliances that are dependent on plastics for electrical insulation.
10. Describe a useful nondestructive test.
11. What type molecular weight data are obtained by osmometry and light scattering techniques?
12. Which polymer solution will exhibit the greater change in osmotic pressure with change in concentration—one in a good or a poor solvent?
13. When a polydisperse polymer is characterized by GPC, which is eluted first—the high or low molecular weight polymer?
14. How can infrared spectroscopy be used to characterize a copolymer?
15. What techniques would you use to differentiate between isotactic and syndiotactic PP?
16. What simple technique could you use to differentiate between a copolymer of styrene and acrylonitrile and one of styrene and methyl methacrylate?

ANSWERS

1. ASTM standards on plastics sponsored by Committee D-20.
2. Failure may occur if a plastic part expands more rapidly when heated than a nonplastic component part in the same machine.

ANSWERS

3. Plastic (b).
4. No. Test procedures must be followed verbatim.
5. Plastic (b).
6. Because they will bend less when subjected to a load.
7. Impact failure.
8. Scratch resistance tests.
9. Electrical cord, circuit breakers, wall receptacles, TV and radio parts, etc.
10. For example, articles such as bottles, telephones or office machine housings may be dropped from a standard height before shipment. Obviously any broken specimens are not shipped.
11. Number average and weight average molecular weight, respectively.
12. A solution of polymer in a good solvent.
13. The largest or highest molecular weight polymers.
14. The intensities of the characteristic absorption bands for specific functional groups are proportional to the concentration of these groups present in the macromolecules.
15. NMR.
16. Pyrolysis G.C.

chapter 7

Phenolic and Other Thermosetting Plastics

7.1 PHENOLIC PLASTICS

That phenol (C_6H_5OH) reacts readily with aldehydes, such as formaldehyde (H_2CO), under both acid and alkaline conditions has been known for over a century. Unfortunately, the early organic chemists, such as Baeyer, Michael, and Kleeberg, used more than 1 mole of formaldehyde per mole of phenol and obtained products that differed from the traditional crystalline products of organic chemistry reactions. These chemists described these resinous masses as goos, gunks, and messes and were quick to abandon the investigation of these reactions and to return to the syntheses of more classical products.

Later attempts by more practical chemists, such as Smith, Luft and Blumer, were more successful since they used lower ratios of formaldehyde to phenol. However, the first commercial phenol-formaldehyde plastic (PF) was not made until Leo Baekeland produced a reproducible thermoplastic that he subsequently converted to a thermosetting plastic by heating the thermoplastic under pressure. The first commercial product, which was produced in the presence of alkalies such as ammonia or sodium carbonate, was patented by Baekeland in 1909. This so-called *resole resin* was made with an excess of formaldehyde as a 40% aqueous solution. The change of the resin from a thermoplastic or *A-stage resin* to a cross-linked *C-stage resin* was controlled by a gradual increase in temperature. It was customary to use 1.5 moles of formaldehyde (45 g) and 1 mole of phenol (94 g).

As shown in Fig. 7-1, the *A*-stage resin is probably a partially polymerized methylolphenol which is fusible and soluble. When this *A*-stage resin is heated under moderate conditions, condensation takes place and a *B*-stage resin is produced. This high molecular weight intermediate is swollen by solvents and softened by heat but is no longer thermoplastic. When the *B*-stage resin is heated it becomes highly crosslinked and is no longer softened by heat nor swollen by solvents. The process during which the resin changes from *A* to *B* to *C*-stages is called *advancement of the resin*. The final *C*-stage is an infusible, insoluble thermosetting plastic.

While the reaction in Fig. 7-1 is shown as a dehydration to produce ether (C—O—C) linkages, the final products actually contain methylene $\left(\begin{array}{c}H_2\\-C-\end{array}\right)$ linkages resulting from a loss of formaldehyde from the ether groups or from the reaction of a methylol group with a benzene hydrogen atom as suggested in Fig. 7-2. In either case, the end result is a substituted diarylmethane. The final crosslinked *C*-stage resin is a three-dimensional network of polymeric chains.

In practice, the water solution of *A*-stage resole resin may be cooled and stored at a low temperature for several weeks. However, these liquid resins have limited shelf life. These aqueous solutions may be used to impregnate paper, wood, or other sheet stock. The impregnated sheets may be plied up as multiple sheets and cured in a hot platen press to produce a *laminate*.

Higher concentrations of *A*-stage resole resins may be obtained by removing some of the water at a low temperature under reduced pressure. The residue may then be dissolved in ethanol and used as a varnish for the impregnation of sheets or other porous forms. A solid is obtained when more water is removed by further evaporation under reduced pressure. The solid product may be ground to a finely-divided resole resin that may be used directly as a binder for wood flour, wood chips, or other particulate matter. Since sufficient formaldehyde to produce a *C*-stage resin is present in the original reactant mixture, these resins are called *one-step resins*.

In a paper presented to the American Chemical Society in 1909, Baekeland also described the procedure used to produce novolac resins by the condensation of limited amounts of formaldehyde with an excess of

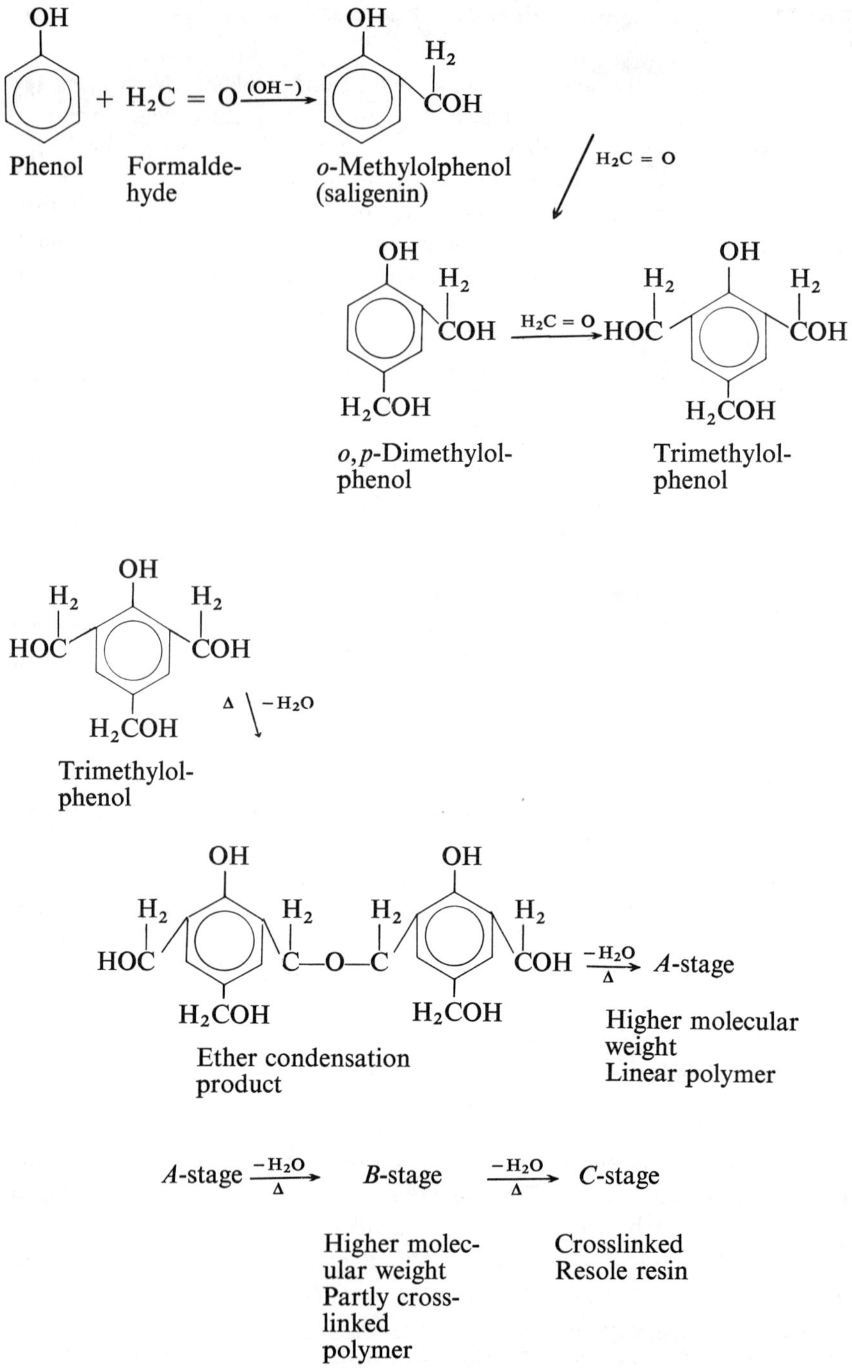

Fig. 7-1 Equations showing possible reactions of phenol and formaldehyde under alkaline conditions.

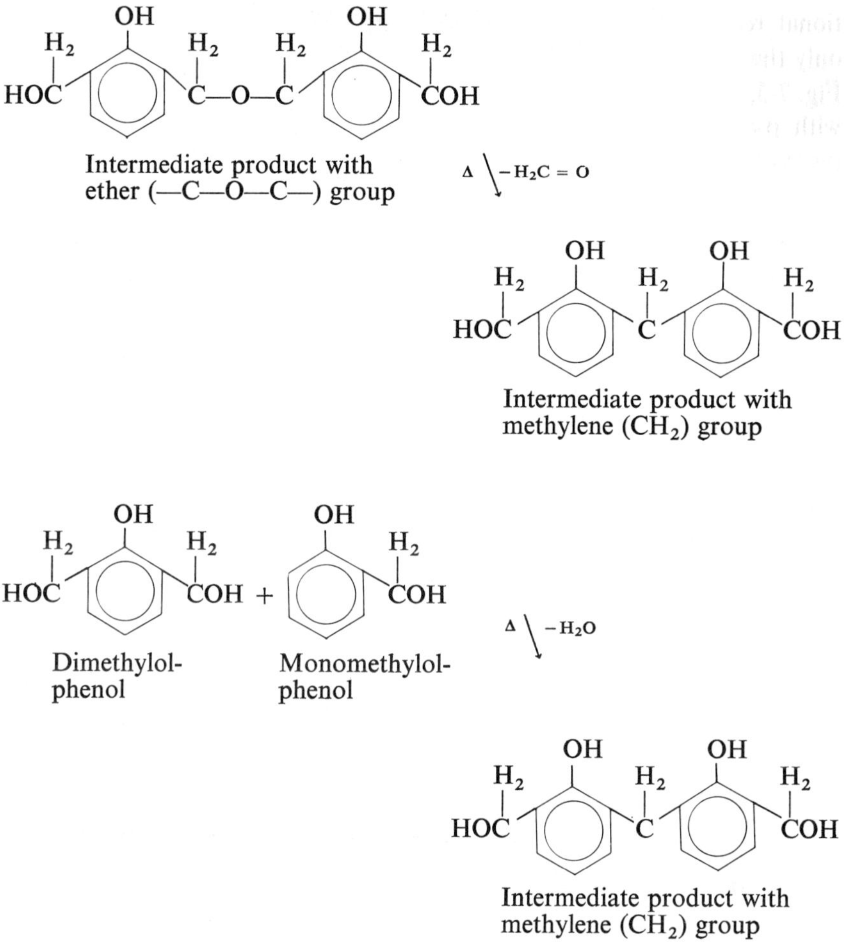

Sec. 7.1   PHENOLIC PLASTICS   143

Intermediate product with ether (—C—O—C—) group

$\Delta \searrow -H_2C=O$

Intermediate product with methylene (CH_2) group

Dimethylol-phenol + Monomethylol-phenol

$\Delta \searrow -H_2O$

Intermediate product with methylene (CH_2) group

Fig. 7-2 Suggested reactions for formation of Methylene (CH_2) groups

phenol in the presence of strong acids. This mechanism, which permits the preparation of an *A*-stage novolac resin, was elucidated by Kienle in 1930. For simplicity, the condensation of p-cresol and formaldehyde will be discussed here in order to demonstrate the relationship of functionality to crosslinking.

Formaldehyde is said to be a difunctional reactant, i.e., another reactant like water or ethanol may possibly attack both the carbon and oxygen atoms in the carbonyl group (C=O) to produce methylene glycol $\left(HOC\overset{H_2}{O}H\right)$

or an acetal [$H_2C(OC_2H_5)_2$]. Likewise, p-cresol also is said to be a difunctional reactant since other reactants such as formaldehyde may react at only the 2 and 6 or ortho positions on the benzene ring. Thus, as shown in Fig. 7-3, a linear thermoplastic is produced when formaldehyde is condensed with p-cresol regardless of the relative proportions of formaldehyde and p-cresol. Since there are only two reactive centers in difunctional reactants, the chain extension can take place in two directions only to produce a linear thermoplastic polymer.

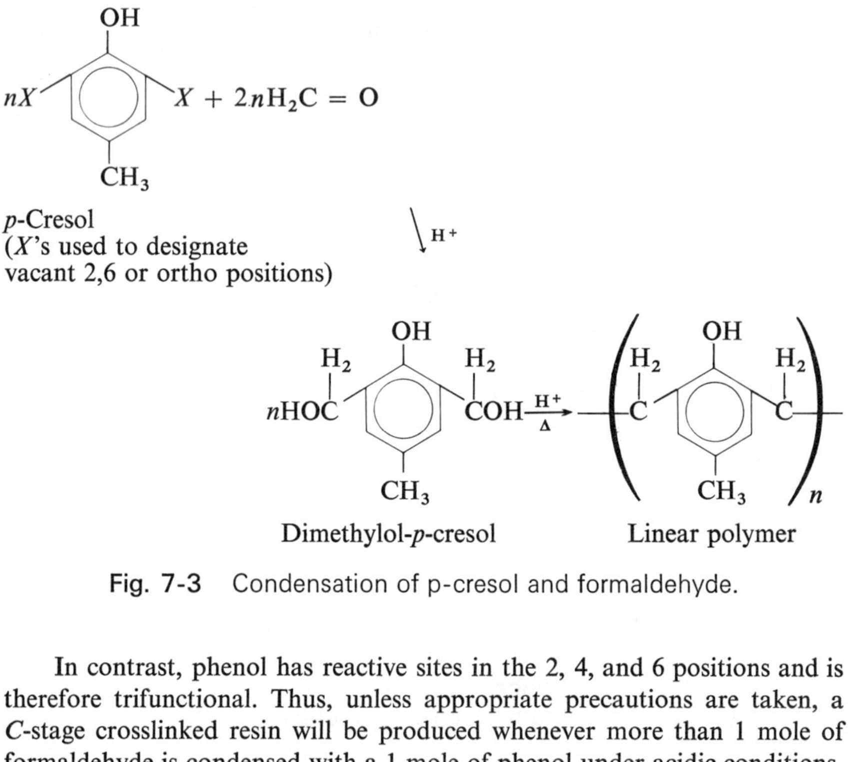

p-Cresol
(*X*'s used to designate vacant 2,6 or ortho positions)

Dimethylol-*p*-cresol   Linear polymer

Fig. 7-3 Condensation of p-cresol and formaldehyde.

In contrast, phenol has reactive sites in the 2, 4, and 6 positions and is therefore trifunctional. Thus, unless appropriate precautions are taken, a *C*-stage crosslinked resin will be produced whenever more than 1 mole of formaldehyde is condensed with a 1 mole of phenol under acidic conditions. The production of a linear *A*-stage resin was assured when Baekeland reacted less than 1 mole of formaldehyde with 1 mole of phenol. Thus, there was not enough of the difunctional formaldehyde present to react with all three centers in the bifunctional phenol.

Condensation is the first step in what is called a *two-step process*, which is characteristic of *novolac resins*. In practice, the formaldehyde-deficient phenol mixture is heated with a small amount of sulfuric or oxalic acid until a viscous solution of a thermoplastic *A*-stage novolac resin is obtained. It is customary to use 0.80 moles of formaldehyde and 1 mole of phenol. The

Sec. 7.1   PHENOLIC PLASTICS   145

water is then removed under reduced pressure and the viscous fusible *A*-stage resin is dumped from the reactor, cooled, and pulverized.

Phenolic molding powders are produced by blending the powdered *A*-stage novolac resin with wood flour or other fillers, pigments, mold lubricants, and a source of additional formaldehyde. The formaldehyde needed to react with the unreacted centers in the trifunctional phenol is supplied by the addition of hexamethylenetetramine. The latter is obtained by the reaction of 6 moles (180 g) of formaldehyde and 4 moles (68 g) of ammonia (NH_3). The type of reaction taking place when this mixture is heated in a mold is shown in Fig. 7-4.

In actual practice, the blend of fusible resin and hexamethylenetetramine is heated on hot differential rolls or in an extruder in order to advance the *A*-stage resin to the *B*-stage. The mixture containing this *B*-stage resin is cooled and pulverized. It is customary to blend this molding powder with

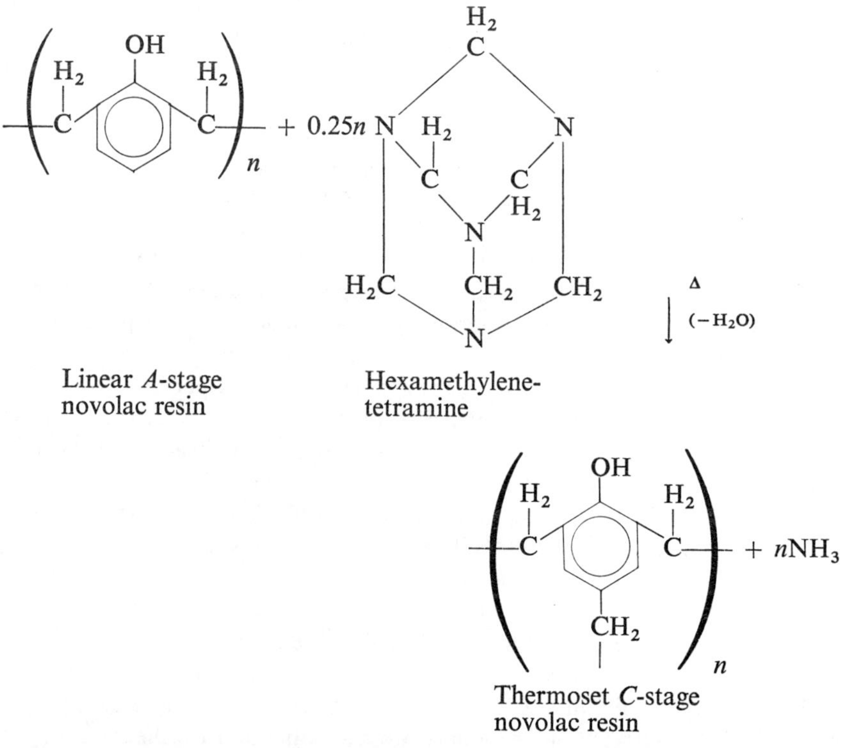

Fig. 7-4 Equation showing possible crosslinking of an *A*-stage resin by hexamethylenetetramine.

other lots of phenolic molding powder to assure uniformity in a large quantity of the molding powder. The presence of fine particles in phenolic molding powders may be eliminated by pelletizing the powder. As shown in Fig. 7-4, some ammonia as well as water is evolved when the molding compound is cured to an infusible plastic in the molding operation.

Wood flour is produced by attrition-type grinding wood waste. It is used as a filler for general-purpose molding powder. The impact resistance is improved by substituting cotton flock, fibrous glass, or asbestos for the wood flour. Asbestos filler contributes to the heat resistance of the end product. The impact resistance of phenolic plastics may be improved by the incorporation of butadiene-acrylonitrile copolymer rubber with the ground *A*-stage resin before the blend is advanced to the *B*-stage.

It is customary to mold phenolic molding compounds at 270–350°F. A typical molded general-purpose phenolic plastic will have the following properties: specific gravity 1.4, tensile strength 7,000 psi, flexural strength 10,000 psi, impact strength 0.3 ft lbs/in. of notch, compressive strength 3,000 psi, Rockwell hardness M110, thermal expansion 4.0×10^{-5} in./in./°C, heat deflection temperature at 264 psi 350°F, good electrical properties, and good resistance to flame, acids, and solvents.

Phenolic resins were the first all-synthetic polymers to be produced commercially. It is of interest to note that they are still being produced using recipes similar to those used by Baekeland seventy years after the pioneer production of these resins. The trade name *Bakelite* has now become a generic name and these plastics are being produced at an annual rate in excess of one billion pounds in the USA.

The resoles are used for laminates and as impregnants for wood and other porous materials. The molded novolac plastics are used for automotive parts, such as distributor caps, for household appliances, such as pot handles, and for electrical parts, such as wall plugs and circuit breakers.

Specialized phenolic resins, which have better flow properties in the mold, have been produced by substituting furfural for formaldehyde. Presumably, this liquid aldehyde, obtained by heating corn cobs with acid, reacts with phenol in a manner similar to that of formaldehyde. Cresols, which are methylphenols, are also used in place of phenol. A thermosetting plastic can not be produced from cresols if the methyl group is present in the 2 or 4 (ortho or para) positions on the benzene ring.

Faster rates of reaction for the production of thermosetting plastics are noted when the methyl or other groups are in the 3 or meta position. Thus, adhesives that cure at relatively-low temperatures are obtained by the reaction of formaldehyde and resorcinol. The latter is a meta-hydroxyphenol. Flexible thermoset resinous products are also obtained by the reaction of formaldehyde and cardinol. The letter is a meta-alkylphenol obtained from cashew nut shells. Sulfonated phenolic resins are used as ion-exchange resins.

7.2 UREA PLASTICS

Reaction products of urea and formaldehyde were described in 1884 and patented by John and Pollack in 1918. However, in spite of the similarity in synthesis to that of phenolic plastics, urea plastics were not made available commercially until 1926 in Great Britain and in 1928 in the USA. Liquid urea-formaldehyde resins, which are used as plywood adhesives, were introduced commercially under the name of "Kaurit" cements in Germany in 1936.

Plastics produced by the reaction of formaldehyde and *melamine* or *cyanuramide* were introduced commercially in 1939. They are often discussed under the term *amino resins*. However, as illustrated in Fig. 7-5, urea contains amide groups $\left(-\underset{\underset{O}{\|}}{C}NH_2\right)$ and thus urea formaldehyde plastics are amide-formaldehyde plastics and not amino plastics. Nevertheless, in its production and sales statistics, the US Tariff Commission usually includes both urea and melamine resins under the heading of amino resins.

Many of the properties of urea plastics are similar to those of the phenolics but, unlike phenolics, the former are not dark in color and are characterized by pastel and translucent colors as well as slightly-superior electrical properties. As evidenced by the four replaceable hydrogen atoms

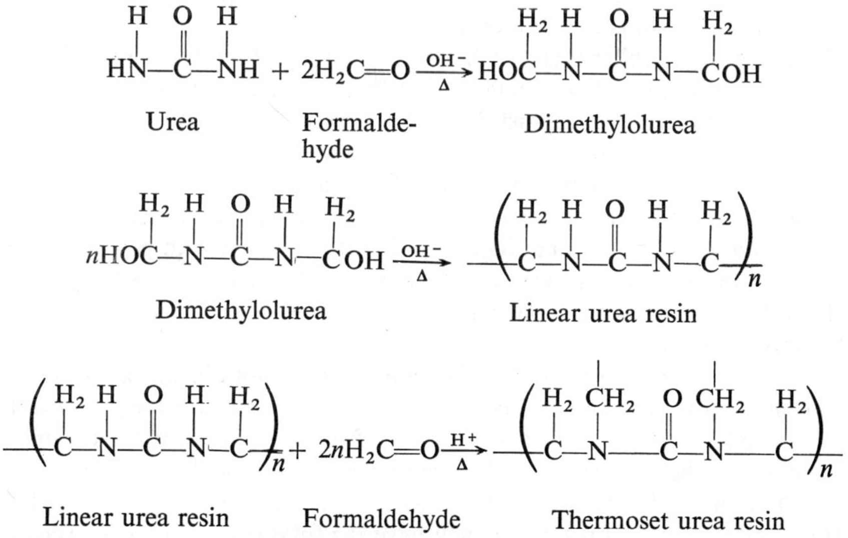

$$\underset{\text{Urea}}{HN-\underset{\|}{\overset{O}{C}}-NH} + \underset{\text{Formaldehyde}}{2H_2C=O} \xrightarrow[\Delta]{OH^-} \underset{\text{Dimethylolurea}}{HOC-N-\underset{\|}{\overset{O}{C}}-N-COH}$$

$$n\underset{\text{Dimethylolurea}}{HOC-N-\underset{\|}{\overset{O}{C}}-N-COH} \xrightarrow[\Delta]{OH^-} \underset{\text{Linear urea resin}}{\left(C-N-\underset{\|}{\overset{O}{C}}-N-C\right)_n}$$

$$\underset{\text{Linear urea resin}}{\left(C-N-\underset{\|}{\overset{O}{C}}-N-C\right)_n} + 2nH_2C=O \xrightarrow[\Delta]{H^+} \underset{\text{Thermoset urea resin}}{\left(C-N-\underset{\|}{\overset{O}{C}}-N-C\right)_n}$$

Fig. 7-5 Reactions of urea and formaldehyde for the preparation of urea resins.

in its formula, urea $\left(\begin{matrix} & O \\ & \parallel \\ H_2N&CNH_2 \end{matrix}\right)$ is tetrafunctional. Nevertheless, it is possible to obtain a thermoplastic resin by heating 1.8 moles of formaldehyde (54 g) with an aqueous solution of 1 mole of urea (60 g) under weak alkaline conditions, such as a pH of 8.5, since urea is difunctional under these conditions.

Figure 7-5 shows that a dimethylolurea is produced under these mild condensation conditions. This product then condenses to a linear polymer. Under acidic conditions, and in the presence of formaldehyde, the latter is converted to a thermosetting plastic in a manner similar to that discussed previously for the curing of phenolic resole resins.

The pioneer urea resins (UF) were produced from a mixture of urea and thiourea $\left(\begin{matrix} & S \\ & \parallel \\ H_2N&CNH_2 \end{matrix}\right)$ under the trade name of *Beetle*. Subsequently, an all urea-formaldehyde plastic called *Plaskon* was developed as a replacement for the metal frames used for scales. Guanidine $\left(\begin{matrix} & H \\ & N \\ & \parallel \\ H_2N-&C-NH_2 \end{matrix}\right)$ and polymethylenediurea $\left(\begin{matrix} O & & & & O \\ \parallel & H & & H & \parallel \\ H_2N-C-&N-(CH_2)_2-&N-C-NH_2 \end{matrix}\right)$ have also been condensed with formaldehyde to produce resins but the principal amide-formaldehyde resin is made from urea.

The principal filler used for the production of light-colored urea plastics is *alpha cellulose*, which has an index of refraction similar to that of the resin. In some instances, a spray-dried powdered resin is compounded with the filler and other compounding ingredients. However, the conventional technique is to admix the alpha cellulose with the aqueous alkaline syrup and to dehydrate this mixture on a moving belt. The product is then ground and a small amount of the ground filled resin is then admixed with pigments in a ball mill in order to produce a master pigment batch. The latter is then blended with the major portion of the ground-filled resin along with lubricants and acid.

The compounded mixture is then advanced on hot differential rolls and reground in the same manner as that described previously for the production of phenolic molding compounds. Wood flour is used in place of alpha cellulose for the production of dark molding compounds.

It is customary to mold urea plastics at 275–350°F. A typical molded alpha cellulose-filled urea plastic will have the following properties: Tensile strength 9,000 psi, impact strength 0.3 ft lbs/in. of notch, flexural strength 14,000 psi, compressive strength 32,000 psi, Rockwell hardness M115, coefficient of thermal expansion 3.0×10^{-5} in./in./°C, heat-deflection temperature at 264 psi, 275°F, good electrical properties, good resistance to flame, and fair resistance to solvents and acids. Unlike phenolic plastic, urea plastic does not carbonize when an electric arc is placed on its surface. It also has high dielectric strength.

Liquid urea resins may be used for the impregnation of paper for the production of decorative laminates, or the liquid may be converted to a fine powder by spray drying. These unfilled resins are widely used as wood adhesives, coatings and textile assistants, such as crease-resistant resins. Since the methylol groups will react with alcohols to produce stable ethers, it is possible to produce linear coating resins by reacting the dimethylolurea with 2 moles of an alcohol such as 1-butanol. Thus, the reaction product is difunctional and will react with an excess of formaldehyde to produce a fusible thermoplastic. A water soluble product is obtained when methanol is used in place of 1-butanol.

Over 50 million pounds of urea molding compounds are used annually for bottle closures, appliance hardware, control knobs, and circuit breakers. The annual production of unfilled urea resins exceeds 200 million pounds. The principal use of these resins is as adhesives, decorative laminates, and textile treating resins. A sizeable quantity of urea resin is also admixed with other thermosetting resins such as alkyds.

7.3 MELAMINE PLASTICS

Since melamine is a 2,4,6-triamino-1,3,5-triazine, it is capable of reacting with formaldehyde to produce an amino resin. This triamine was isolated by Liebig in 1834, but prior to the 1930s it was an expensive laboratory chemical. Other related amino compounds such as benzoguanamine and its derivatives and aniline will also react with formaldehyde to produce amino resins, but the principal amino resin is produced by the condensation of melamine and formaldehyde.

These resins were patented in England in 1935. Subsequently, several American firms applied for patents for these resins. The basic US patent was granted in 1939 to the company whose research chemist kept the best records.

As shown in Fig. 7-6, melamine has 3 amino groups, each of which is capable of reacting with two molecules of formaldehyde. Thus, it is possible to produce hexamethylolmelamine by the alkaline condensation of 6 moles of formaldehyde (180 g) with 1 mole of melamine (126 g). As was the case with methylol urea, this hexadroxy compound will react with formaldehyde to form hexamethoxymethylmelamine. The melamine-formaldehyde resins (MF) are stable under alkaline conditions like urea formaldehyde resins, but cure to thermosetting plastics when an acid is added. It is customary to add enough formaldehyde to produce only two or three methylol groups on each melamine molecule.

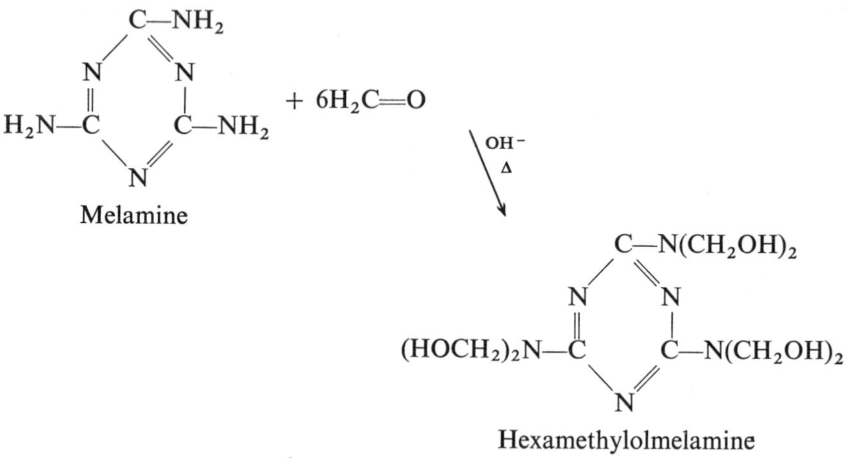

$$\text{Melamine} + 6H_2C=O \xrightarrow[\Delta]{OH^-} \text{Hexamethylolmelamine}$$

Di or trimethylolmelamine $\xrightarrow[\Delta]{H^+}$ Thermoset melamine plastic

Fig. 7-6 Reaction of melamine and formaldehyde to produce melamine plastics.

The production techniques and the volume of plastics used is similar for urea and melamine plastics. The latter are more expensive but have better heat resistance than urea plastics. Thus it is customary to mold melamine plastics at a slightly-higher temperature than that used for urea plastics.

Molding powders are produced by mixing alpha cellulose or some other filler such as asbestos with the aqueous alkaline solution of the product obtained by the condensation of formaldehyde and melamine. Wood flour, asbestos, and fibrous glass are utilized as fillers for the production of plastics used for industrial applications. One of the most widely-used applications of melamine plastics and one of the best known case histories of sound marketing for plastics is the molding and distribution of melamine plastic dinnerware. The patterns on these articles and on decorative laminates is produced by placing a printed sheet impregnated with a solution of melamine resin on top of the molded pieces and repeating the molding operation.

A typical alpha cellulose-filled melamine plastic will have the following properties: Tensile strength 10,000 psi, flexural strength 13,000 psi, impact strength 0.3 ft lbs/in. of notch, compressive strength 42,000 psi, Rockwell hardness M120, coefficient of thermal expansion 4.0×10^{-5} in./in./°C, heat-deflection temperature at 264 psi, 360°F, very good electrical properties, good resistance to flame, and fair resistant to solvents and acids. Neither the melamine, urea, or phenolic plastics are recommended for continuous exposure to strong alkalies.

7.4 ALKYD PLASTICS

The first polyester resin was synthesized by Berzelius in 1847 by the condensation of glycerol and tartaric acid. The first commercial product called *glyptal resin* was produced in the early part of the 20th century by the controlled condensation of glycerol and phthalic acid. The pioneer resins were used as coatings in which the principles of functionality were of utmost importance. As illustrated in Fig. 7-7, these polyester resins were prepared by the acid condensation of a difunctional acid viz., phthalic acid or its corresponding anhydride and glycerol. The primary hydroxyl groups on the terminal carbon atoms of glycerol are more reactive than the secondary hydroxyl group on carbon 2.

Thus, as shown in Fig. 7-7, glycerol acts as a difunctional reactant at moderate temperatures and thus yields a linear polyester containing a reactive hydroxyl group.

The unreacted secondary hydroxyl groups in the linear polyester will form esters with phthalic acid at elevated temperatures. Thus, the linear polyester will cure when heated. The last step usually takes place after the linear resin has been applied to an object as a coating or is present in its final shape.

In the early 1930s, Kienle coined the term *alkyd* from parts of the names of the reactant *al*-cohol and a-*cid*. The term is often applied to any polyester resin but is generally used to describe a polyester with unsaturated groups in its linear intermediate. The paint industry obtains this lack of saturation by the addition of unsaturated monocarboxylic acids such as oleic, linoleic, linoleic acid, or esters of these unsaturated acids. Actually, the widely-used terms of *short, medium,* and *long alkyds* are used to denote an increase in the relative amount of unsaturated acid used.

A curable alkyd resin can be produced by the condensation of a difunctional acid and alcohol such as phthalic acid and ethylene glycol ($HO(CH_2)_2OH$) in the presence of an unsaturated acid or ester. Unsaturated linear polyesters of this type, called *prepolymers*, are mixed with fillers such as clay or ground limestone and peroxy compounds. These *premixes* are sometimes used as putties, called *bulk-molding compounds* (BMC), or as impregnated sheets, called *sheet-molding compounds* (SMC). Unsaturated polyester resins, which are discussed in Sec. 7.5, are also used in BMC and SMC formulations.

Alkyd resins are produced at an annual rate of over 600 million pounds. Most of this production is used in the paint and coatings industry, but over 50 million pounds are used annually for molding various articles.

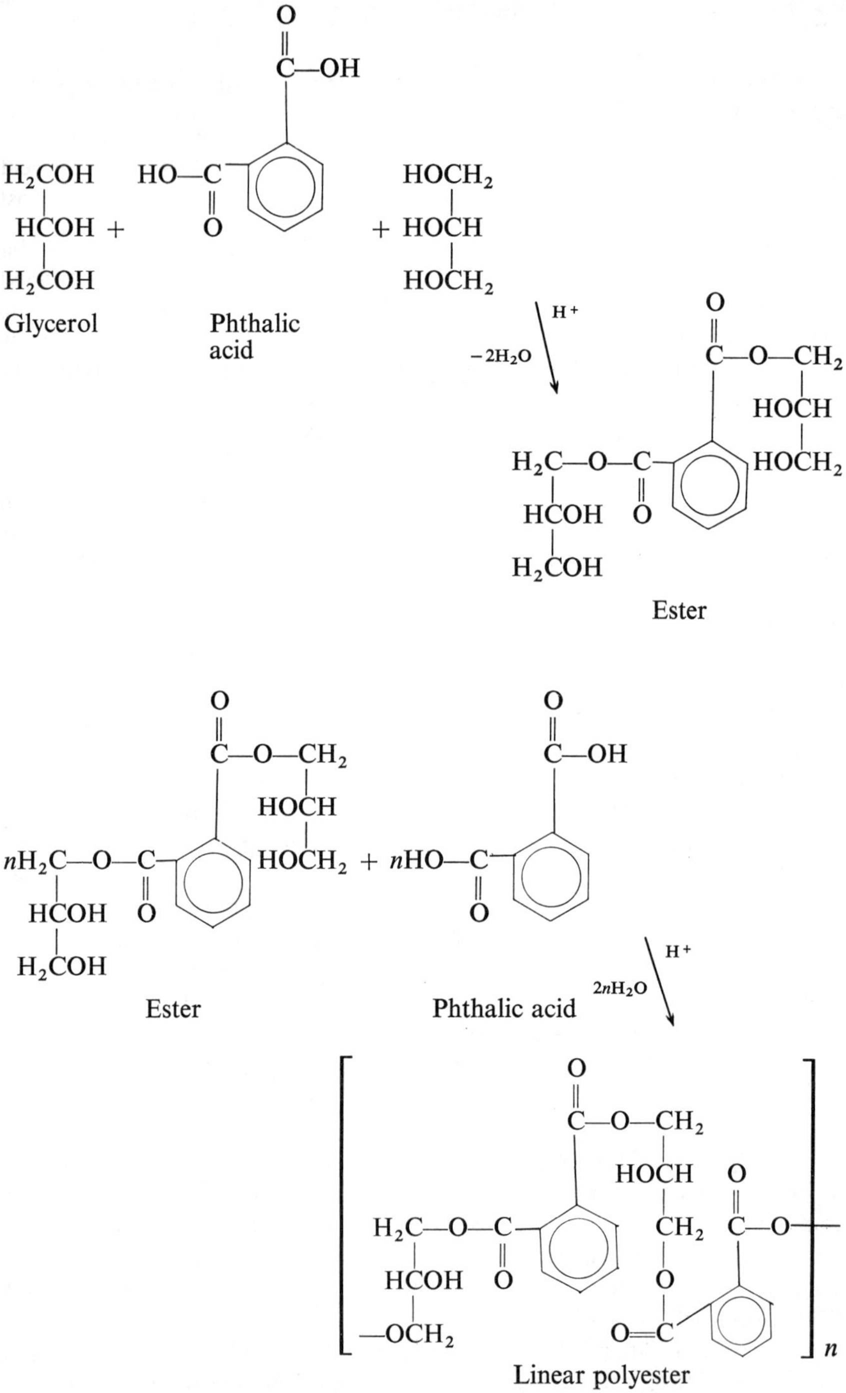

Fig. 7-7 Equation for the reaction of glycerol and phthalic acid at moderate temperatures.

A typical cured fibrous glass filled alkyd plastic will have the following properties: Tensile strength 7,000 psi, flexural strength 7,000 psi, impact strength 10 ft lbs/in. of notch, compressive strength 30,000 psi, coefficient of thermal expansion 20×10^{-6} in./in./°C, Barcol hardness 75, heat deflection temperature at 264 psi, 400°F, good electrical properties, and fair resistance to flame, solvents, and acids.

7.5 UNSATURATED POLYESTER PLASTICS

In 1940, Ellis patented the production of unsaturated polyester resins, differing from alkyds in that the unsaturation was the result of the use of an unsaturated dibasic acid viz., maleic acid, or its anhydride. This type of unsaturated polyester is compatible with liquid vinyl monomers such as styrene and the mixture can be cured in the presence of peroxy compounds to produce insoluble crosslinked products, as shown in Fig. 7-8.

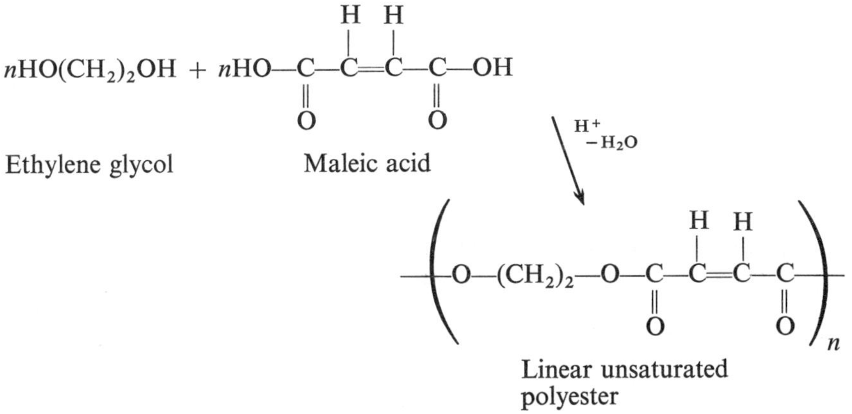

$$n\text{HO(CH}_2)_2\text{OH} + n\text{HO}-\underset{\underset{O}{\|}}{C}-\underset{H}{\overset{H}{|}}{C}=\underset{H}{\overset{H}{|}}{C}-\underset{\underset{O}{\|}}{C}-\text{OH} \xrightarrow[-\text{H}_2\text{O}]{\text{H}^+}$$

Ethylene glycol   Maleic acid

$$\left(-\text{O}-(\text{CH}_2)_2-\text{O}-\underset{\underset{O}{\|}}{C}-\underset{H}{\overset{H}{|}}{C}=\underset{H}{\overset{H}{|}}{C}-\underset{\underset{O}{\|}}{C}- \right)_n$$

Linear unsaturated polyester

Fig. 7-8 Equation for the synthesis of a linear unsaturated polyester.

It is customary to add metallic soaps and tertiary amines to accelerate the polymerization and to incorporate thixotropic agents such as pyrogenic or fumed silica to control the flow. Fillers, such as clay, talc, and alumina, are usually added to the resin mix to reduce shrinkage during curing. Fibrous reinforcements, such as asbestos, fibrous glass, or poly(vinyl alcohol) fibers, are usually added to produce strong plastics.

Polyester-molded products are essentially colorless and may be pigmented to produce any color desired in the end product. Polyester plastics are used at the rate of over 600 million pounds a year for the production of panels, roofing, tanks, profiles, boats, and many structures.

The properties of a typical molded fibrous glass-filled polyester are as follows: Tensile strength 7,000 psi, flexural strength 13,000 psi, compressive strength 25,000 psi, impact strength 10 ft lbs/in. of notch, Barcol hardness 65, coefficient of thermal expansion 2.5×10^{-5} in./in./°C, heat deflection temperature at 264 psi, 400°F, very good electrical properties, and fair resistance to solvents and acids.

Over one billion pounds of reinforced plastics are produced annually in the USA. The principal reinforcing agent is fibrous glass, and this is usually extended with clay, talc, or alumina. The principal resin is an unsaturated polyester dissolved in styrene but this may be replaced by vinyltoluene, methyl methacrylate, or diallyl phthalate. The linear unsaturated polyester resin is usually a condensate of ethylene glycol, shown in Fig. 7-8, but this may be replaced at least in part by propylene glycol, dimethylene glycol, dipropylene glycol, and bisphenol A.

The unsaturated dicarboxylic acid is maleic acid or its anhydride, but this may be replaced by fumaric acid. It is also customary to reduce the extent of unsaturation and subsequent crosslinking by blending these unsaturated acids with phthalic anhydride, isophthalic acid, and adipic acid.

It is customary to inhibit the monomer-polyester solution during storage by the addition of tert.-butylcatechol or hydroquinone.

The so-called catalysts used to cure unsaturated polyester resins are actually initiators, such as methyl ethyl ketone peroxide, which, as stated in Sec. 2.2, supply free radicals that become part of the polymer chain. Other initiators used for polyester resin curing at moderate temperatures are cyclohexanone peroxide, and benzoyl peroxide. Initiators used at higher temperatures are cumene hydroperoxide, tert.-butyl per benzoate, and tert.-butyl peroctoate.

These compositions may be cast, laid up by hand, sprayed, or molded. The final products are used as corrugated or flat panels, helmets, electrical appliances, furniture, containers, automotive housings, boats, house sidings, modular bathrooms, and chemical storage tanks.

7.6 EPOXY RESINS

Epoxy (EP) or epoxide resins are characterized by the presence of a reactive oxirane, ethoxylene, or epoxy group $\left(\begin{array}{cc} H & H \\ | & | \\ -C-C- \\ \diagdown O \diagup \end{array}\right)$ in the backbone of the uncured resinous intermediate. Glycidyl ethers were prepared by the condensation of epichlorohydrin and monofunctional phenol by Lindermann

Sec. 7.6 EPOXY RESINS

in the last part of the nineteenth century. He produced an ether with a difunctional phenol but, unfortunately, he chose catechol which produced a cyclic ether and not a linear diglycidyl ether. However, Schlack described such ethers, including that prepared from bisphenol A, in a German patent in 1934. The first patent on epoxy resins was a Swiss patent issued to Castan in 1940.

The original epoxy resins were sold under the trade name Araldite. Many improvements in formulation were patented by Greenlee and assigned to Devoe and Reynolds in the late 1940s and early 1950s. The trade names Epon and Epikote were adopted by Shell for sale of these resins in the USA and Europe, respectively.

The classical intermediate, and the one accounting for 85% of all production of epoxy resins, is a diglycidyl ether produced by the gradual addition of sodium hydroxide to promote the condensation of more than 2 moles of epichlorohydrin and 1 mole of bisphenol A [2,2'-bis(4-hydroxyphenyl)propane] at 60°C. Lighter-colored resins are produced when zinc chloride or lithium compounds are used instead of sodium hydroxide as the condensation catalyst. As depicted in Fig. 7-9, the molecular weight of this intermediate may be increased by further reaction with bisphenol A.

Whether the intermediate is a liquid or solid depends on the extent of the second reaction shown in Fig. 7-9. When the value of n is small, such as 3 or 4, the intermediate is liquid and has a large number of reactive epoxy groups per unit weight. When the value of n is large, such as $n = 20$, the intermediate is a solid and has fewer number of reactive epoxy groups per unit weight.

Another type of epoxy resin intermediate is the *epoxidized novolac resins* obtained by the alkaline condensation of epichlorohydrin with an *A*-stage phenolic-formaldehyde-novolac resin, such as that discussed previously in Sec. 7.1. Because of the presence of more hydroxy groups in these methylolphenols, the epoxidized novolac intermediates have more epoxy groups, i.e., they have a higher functionality than the products obtained from bisphenol A. These intermediates are preferred for the production of epoxy resins for high temperature applications.

Another intermediate with high functionality is produced by the alkaline condensation of epichlorohydrin with p-aminophenol $\left(H_2N-\hspace{-4pt}\bigcirc\hspace{-4pt}-OH\right)$.

Since both the hydrogen atoms on the amine group as well as the phenolic hydrogen react with epichlorohydrin, the simplest intermediate contains three epoxy groups. These intermediates cure rapidly at moderate temperatures to produce resins with excellent adhesive properties.

The pioneer curing-agent used by Castan was phthalic anhydride which crosslinked with the hydroxyl groups shown in the structure in Fig. 7-9.

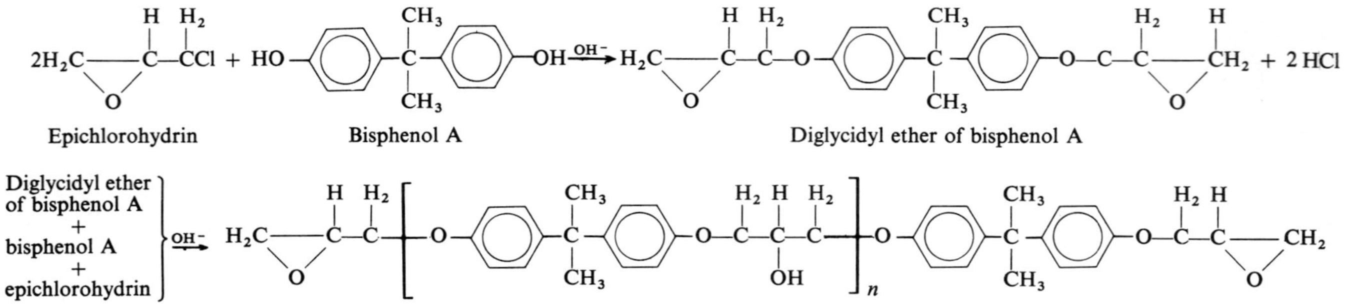

Fig. 7-9 Equation for the preparation of epoxy resin intermediates.

Fig. 7-10 Structures of bis-(2,3-epoxycyclopentyl) ether and a cyloaliphatic epoxy carboxylate respectively.

Any partly-cured products were termed *structopendant prepolymers*. In his U.S. patent in 1948, Castan also included the use of amine curing agents which cause crosslinking through the epoxy end groups and produce *structoterminal prepolymers*.

The reaction of terminal epoxy groups with a polymethyleneamine is a general reaction that takes place at moderate temperatures. Hence, almost any compound such as ethers, esters, or amines containing two or more epoxy groups may be used to produce epoxy resins.

In addition to the intermediates cited previously, bis-(2,3-epoxycyclopentyl) ether and cycloaliphatic epoxy carboxylates have been used. The former produces strong composites for aerospace applications. Representative structures for these epoxy compounds are shown in Fig. 7-10.

Any reactive polyfunctional epoxy-reactive compound may be used to convert an epoxy resin intermediate to a hard plastic. It is customary to use polymethyleneamines for the cure or hardening of bisphenol A epoxy intermediates at moderate temperatures. These hardeners are sometimes called *catalysts*, but since they react with the epoxy groups to produce polymer in which the hardeners are combined chemically in the polymer chain, they are reactants and not catalysts. The curing of these epoxy resin intermediates is sometimes called a *condensation reaction*, but this is also incorrect since the resinous product is produced exclusively in the absence of any byproducts. The term step reaction polymerization is preferred.

Aliphatic polyamines, such as diethylenetriamine, triethylenetetramine, tetraethylenepentamine and diethylaminopropylamine, have been used as hardeners to produce crosslinked epoxy resins at moderate temperatures. Aromatic amines, such as meta-phenylenediamine and methylenedianiline, produce epoxy resins with better heat resistance than the aliphatic amines, but when used as curing agents, the aromatic amines require higher temperatures than the aliphatic amines.

Cycloaliphatic epoxide intermediates must be cured at elevated temperatures with polycarboxylic acids or cyclic anhydrides that react with the hydroxyl groups in these intermediates. Any epoxy resin intermediate cured with a cyclic anhydride, such as hexahydrophthalic anhydride, will have better resistance to elevated temperatures than amine-cured resins. Resinous products with still higher heat-deflection temperatures are obtained when pyromellitic dianhydride and chlorendic anhydride are used as hardeners.

Epoxy resins are used as coatings, potting compositions, adhesives, and encapsulations as well as for molded and reinforced plastics. Over 150 million pounds of epoxy resins were produced in the USA in 1972. The properties of a typical molded fibrous glass-filled epoxy plastic are as follows: Tensile strength 10,000 psi. flexural strength 14,000 psi, compressive strength 24,000 psi, impact strength 1.0 ft lbs/in. notch, Rockwell hardness M100, coefficient of thermal expansion 4.0 × 10^{-5} in./in./°C, heat deflection temperature at 264 psi, 300°F, good electrical properties, and good resistance to solvents and acids.

7.7 ALLYL PLASTICS

The term allyl resins and plastics is used to describe products obtained by the polymerization of allyl monomers containing more than one allyl group. The pioneer representative of this class was a cast polymer of diethylene glycol bis(allyl carbonate) that was introduced in 1941 and is sold under the trade name CR-39. Because of ease of grinding and good optical properties, castings obtained by the peroxide-initiated polymerization of this monomer in situ, have been used for lenses in eye glasses.

Another polyallyl monomer used to produce polymers by casting in situ is triallyl cyanurate (TAC). This polymer has unusually good resistance to elevated temperature. In some cases, the monomer has been admixed with other vinyl or diallyl monomers to improve the high temperature properties of other plastics.

The most widely used allyl plastics are produced by the polymerization of diallyl phthalate or diallyl isophthalate. These are used as coatings, as impregnant for decorative laminates, as molding powders, and as additives to vinyl monomers to promote crosslinking. Diallyl maleate and diallyl chlorendate have also been used to a limited extent.

It is customary to polymerize diallyl phthalate with an organic peroxide initiator and to stop the polymerization while the polymer is still in the linear or *B*-stage. This prepolymer is then blended with more monomer, initiator, and filler, and advanced using techniques similar to those described previously for phenolic molding powder. The structural formulas for these allyl monomers are shown in Fig. 7-11.

A typical molded fibrous glass reinforced allyl plastic (Dapon) has the following properties: Tensile strength 8,000 psi, flexural strength 15,000 psi, compressive strength 30,000 psi, impact strength 0.6 ft lbs/in. of notch, Rockwell hardness M105, coefficient of thermal expansion 3.0 × 10^{-5} in./in./°C, heat deflection temperature at 246 psi, 350°F, excellent electrical properties, and good resistance to solvents and acids.

7.8 SILICONES

Since silicon resembles carbon, silicon atoms can catenate to some extent to form silanes $(H(SiH_2)_nH)$ which are counterparts of the corresponding carbon containing alkanes $(H(CH_2)_nH)$. Unlike the alkanes, which are made

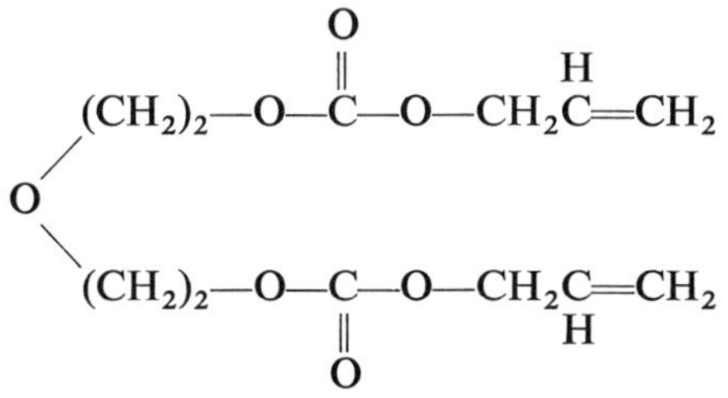

$$\begin{array}{l}\diagdown\text{(CH}_2)_2-\text{O}-\overset{\overset{\text{O}}{\|}}{\text{C}}-\text{O}-\text{CH}_2\overset{\text{H}}{\text{C}}=\text{CH}_2\\ \text{O}\diagup\\ \diagdown\text{(CH}_2)_2-\text{O}-\underset{\underset{\text{O}}{\|}}{\text{C}}-\text{O}-\text{CH}_2\underset{\text{H}}{\text{C}}=\text{CH}_2\end{array}$$

Diethylene glycol bis (allyl carbonate)

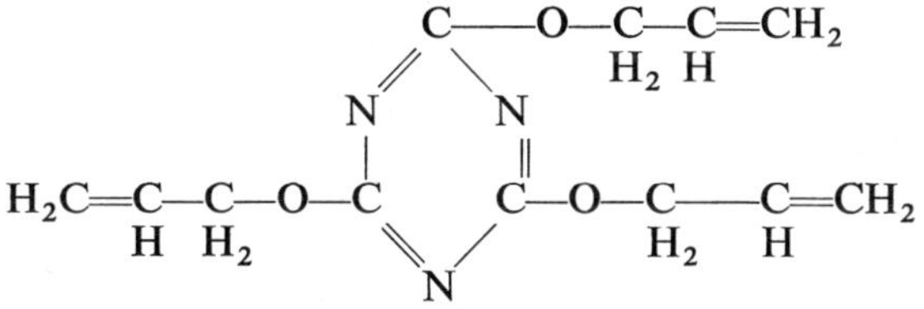

Triallyl cyanurate

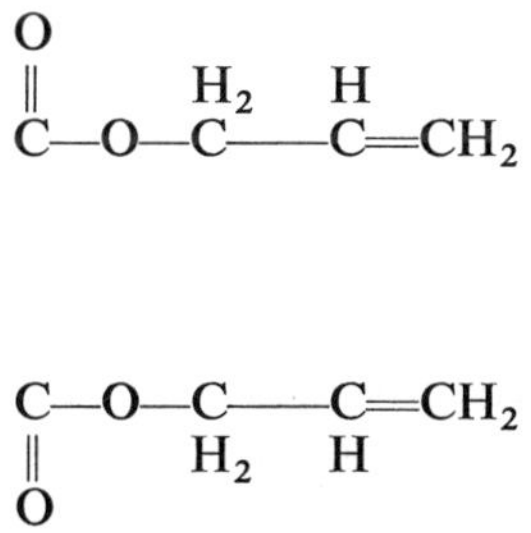

Diallyl phthalate

Fig. 7-11 Structural formulas for allyl monomers.

up of relatively strong carbon-carbon bonds, the silicon-silicon bonds are weak. Siloxane bonds (—O—Si—) are stronger and greater stability is noted when alkyl groups are present on the silicon atoms in siloxanes. The latter were originally thought to be ketones and were called silicones by mistake.

The classical investigations of these organosilicon products, which are still called silicones, were undertaken in the early part of the nineteenth century by Kipping. This investigator like so many other research chemists failed to consider the practical aspects of his research. Commercial silicones were developed by Hyde, McGregor, and Rockow in the 1930s and 1940s.

In the classical Kipping synthesis, alkyl chlorosilanes, such as dimethyldichlorosilanes (($H_3C)_2SiCl_2$) are prepared by the reaction of methylmagnesium chloride and silicon tetrachloride ($SiCl_4$) in a stepwise reaction. This difunctional reactant may be reacted further to produce trimethylmonochlorosilane (($H_3C)_3SiCl$), which is monofunctional.

A linear polymer is obtained when the dimethyldichlorosilane is hydrolyzed, as shown in Fig. 7-12.

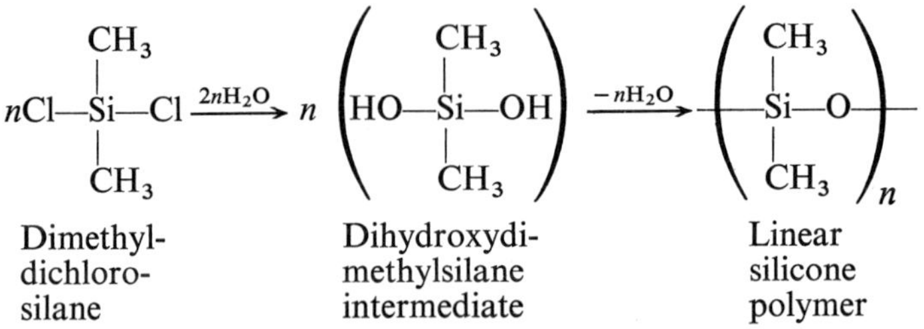

$$n\text{Cl}-\underset{\underset{CH_3}{|}}{\overset{\overset{CH_3}{|}}{Si}}-\text{Cl} \xrightarrow{2n H_2O} n\left(\text{HO}-\underset{\underset{CH_3}{|}}{\overset{\overset{CH_3}{|}}{Si}}-\text{OH}\right) \xrightarrow{-nH_2O} \left(\underset{\underset{CH_3}{|}}{\overset{\overset{CH_3}{|}}{Si}}-\text{O}\right)_n$$

Dimethyl- Dihydroxydi- Linear
dichloro- methylsilane silicone
silane intermediate polymer

Fig. 7-12 Synthesis of silicones.

Low molecular weight linear silicones have been used as additives for polishes, lubricants, release agents, and as water repellants. Silicone resins are prepared by the hydrolysis in situ of a mixture of dimethyldichlorosilane and monomethyltrichlorosilane (CH_3SiCl_3). The latter is trifunctional and serves as a crosslinking agent.

Silicone resins are usually reinforced with fibrous glass, asbestos, mica or silica. These products are used for potting, encapsulating, molding, and for the formation of laminates. These plastics are characterized by good resistance to elevated temperatures, resistance to moisture and some chemicals, and excellent electrical properties.

A typical fibrous glass filled molded silicone will have the following properties: Tensile strength 5,000 psi, flexural strength 12,000 psi, compressive strength 12,000 psi, impact strength 10 ft lbs/in. of notch, Rockwell hardness M84, coefficient of thermal expansion 8×10^{-6} in./in./°C, heat deflection temperature at 66 psi 900°F, excellent electrical properties, and fair resistance to solvents and to acids.

BEHAVIORAL OBJECTIVES

Because of their resistance to heat and solvents, thermosetting plastics are an important class of products in the plastic industry. The principal thermosetting plastic is the phenol-formaldehyde polymer which is sometimes called the workhorse of the plastic industry.

After studying this chapter, you should understand the following concepts:

1. Resole resins are thermoplastic phenol-formaldehyde one-step polymers which advance to thermoset plastics when heated.
2. Novolac resins are more stable than resole resins since the A-stage resin is a linear polymer which is made with insufficient formaldehyde so that no crosslinking occurs until formaldehyde is released in the mold by heating with hexamethylenetetramine.
3. The amino resins are essentially resoles since they are readily converted to infusible plastics when heated in the presence of acids.
4. Alkyd resins are unsaturated polyesters produced by the condensation of a dicarboxylic acid and a dihydric alcohol in the presence of an unsaturated monocarboxylic acid. The prepolymers may be cured by heating.
5. Another type of unsaturated polyester is produced by the condensation of a dihydric alcohol and an unsaturated dicarboxylic acid such as maleic acid. These polyesters are usually dissolved in styrene and cured with an initiator such as benzoyl peroxide.
6. Epoxy resins contain epoxy groups which may react with a polyamine to produce a crosslinked polymer.
7. Epoxy resins also contain hydroxyl groups which may react with phthalic anhydride when heated to produce a crosslinked polymer.
8. Crosslinked allyl plastics may be produced by heating a solution of a diallyl ester prepolymer with the diallyl ester monomer in the presence of an initiator such as benzoyl peroxide.
9. Silicone molecules in which both silicon (Si) and oxygen atoms are joined together in the polymeric chain are heat resistant because of the strong siloxane bonds (Si—O—Si).
10. Silicones are water repellant and are good lubricants because of the alkyl (R) pendant groups on the silicon atoms in the polymeric molecule.

GLOSSARY

A-stage: a linear prepolymer of phenol and formaldehyde.
Acid anhydride: a compound which yields an acid when water is added.
Aliphatic: a continuous hydrocarbon chain, so-called because it is oil-like.
Alkyd: an unsaturated polyester resin.
Alkyl: $\cdot(CH_2)_n H$ such as $\cdot CH_3$.
Alpha cellulose: fibrous cellulose insoluble in 17.5% aqueous sodium hydroxide.
Amine group: $—NH_2$.
Amino resins: a class of thermosetting resins which includes urea and melamine resins.

Baekeland, Leo: the inventor of phenol-formaldehyde polymers.
BMC: bulk molding compound.
Bisphenol A: a bifunctional derivative of phenol.
Cardinol: a substituted phenol obtained from cashew nut shells.
Catalyst: a substance which accelerates a reaction and may be recovered unchanged after the reaction has taken place.
Cresol: a methyl substituted phenol.
DAP: diallyl phthalate.
Difunctional: a molecule with two reactive groups.
Drier: a salt of an organic acid and a heavy metal such as cobalt which catalyzes polymerization in the presence of oxygen.
Hexa: hexamethylenetetramine, i.e., the reaction product of formaldehyde and ammonia.
Long oil length alkyd: a highly unsaturated alkyd resin.
MF: melamine-formaldehyde plastics.
Molding powder: a mixture of resin, filler, pigment, and other additives required for the production of a molded plastic.
Novolac: a two-step PF resin produced by the acid condensation of phenol and a controlled amount of formaldehyde. This A-stage resin is then advanced to an infusible plastic by reaction with additional formaldehyde.
Pendant group: a group attached to the polymer chain or backbone of a macromolecule.
PF: phenolic plastic.
Resole: a one-step PF resin made by the alkaline condensation of phenol with adequate amounts of formaldehyde.
Resorcinol: a 3-hydroxyphenol.
Short oil length alkyd: one with few unsaturated groups.
Siloxane: —O—Si—
SMC: sheet molding compound.
TAC: triallyl cyanurate.
UF: urea-formaldehyde plastics.

Unsaturated: an organic compound having an ethylenic group $\left(\begin{array}{c} \diagdown \\ \diagup \end{array} C = C \begin{array}{c} \diagup \\ \diagdown \end{array} \right)$

Wood flour: finely divided fibrous wood.

SUGGESTED QUESTIONS

1. What is the principal difference between an A-stage and a C-stage resin?
2. Which has the longer shelf life before being converted to a thermoset plastic—a resole or a novolac resin?
3. When would you specify UP in place of PF?

4. When would you specify MF in place of UF?
5. Which type of alkyd resin will form the most crosslinks when heated?
6. What type of reactant would you use to cure an epoxy resin at room temperature?
7. What type of reactant would you use to cure an epoxy resin at elevated temperatures?
8. How many unsaturated groups or ethylenic groups are present in diallyl phthalate?
9. What atoms besides silicon are present in the backbone of the silicone molecule?
10. Why are silicones water repellant?

ANSWERS

1. The A-stage is a linear thermoplastic resin and the C-stage is an infusible thermoset plastic.
2. Novolac resin.
3. When light colored products are required.
4. When resistance to heat and moisture is required.
5. A long oil alkyd.
6. A polyamine.
7. An anhydride of a dicarboxylic acid.
8. Two.
9. Oxygen.
10. Because the outside of the macromolecular chain consists of aliphatic hydrocarbon groups which are nonpolar and insoluble in water.

chapter 8

Polyolefins

8.1 LOW DENSITY POLYETHYLENES

Polyethylene originally was called *polymethylene* since it first was produced in the laboratory by Von Peckmann in 1898 from diazomethane. This classical product was actually a high molecular weight crystalline linear polymer of methylene (—CH_2—). Carothers produced a comparable product in 1930 by heating α,ω-dibromoalkanes, $Br(CH_2)_nBr$ with sodium. Paraffin wax is, of course, a low molecular weight polymethylene, but its molecular weight is far below that of the plastic range.

The accidental or serendipitous discovery of a polyethylene that could be produced economically was based on some high pressure studies of ethylene (H_2C=CH_2) by Michaels at Amsterdam. This high pressure technique was used in 1933 by Gibson to produce polyethylene from a mixture of ethylene and benzaldehyde in the presence of a trace of air.

Sec. 8.1 LOW DENSITY POLYETHYLENES

Penn and Williams polymerized ethylene itself at 30,000 psi in 1935, and a British patent which was filed in 1936 was issued to Faucett, Gibson, and others in 1937 for the production of this polymer, called *Polythene*.

The product from a small commercial plant built in Great Britain in 1939 proved to be the best high-insulating material available for coaxial cable for radar applications. Hence, many production facilities were built during World War II. The first American plant was built in 1943 using know-how loaned by Imperial Chemical Industries Ltd. (ICI) in England. This high-pressure process still is being used today to produce over three million tons of this type polyethylene annually in the USA.

The product was originally called *high-pressure polyethylene*, but now, in accord with ASTM standard nomenclature, a polymer of ethylene having a density of 0.910–0.925 g/cm^3 is called *low-density polyethylene* (LDPE). This polymer has a high volume and thus a low density because of the many branches present on the polymer backbone. The extent of branching and the degree of crystallinity increase as the reaction pressure is increased. The degree of crystallinity in commercial LDPE is 60–70%. The opacity due to the presence of crystals is reduced by decreasing the size of the spherulites or clusters of crystals.

The original polymer was made by the high pressure polymerization of ethylene at 475–570°F in the presence of 0.05% oxygen as an initiator. It is now customary to conduct this polymerization in the presence of a peroxy initiator such as benzoyl peroxide in an autoclave or jacketed tube at a pressure of 30,000–40,000 psi. A pressure of 9,000 psi is considered to be the minimum usable pressure. In some instances, this polymerization is conducted in the presence of benzene or chlorobenzene and an inert gas may be present as a chain stopper. This exothermic reaction is terminated when approximately 25% of the ethylene has been converted to polymer. The equation for this reaction is shown in Fig. 8-1.

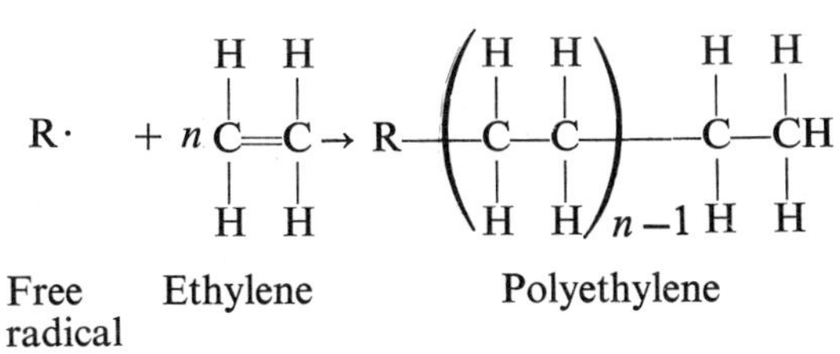

$$R\cdot + n\,C=C \rightarrow R-\left(\begin{array}{c}H\ H\\|\ \ |\\C-C\\|\ \ |\\H\ H\end{array}\right)_{n-1}\!\!\!\!-C-CH$$

Free radical   Ethylene   Polyethylene

Fig. 8-1 Equation for the polymerization of ethylene.

A very low molecular weight polyethylene is produced by the polymerization of ethylene in benzene in the presence of larger amounts of initiator at high pressures. This product has an average molecular weight of less than

10,000. It is used in place of other waxes for making candles, polishes, inks, and coatings.

It is customary to add antiblock agents to LDPE to prevent film from sticking and to reduce the tendency of molded parts to gather dust. Carbon black is often added to protect the polymer against the effects of ultraviolet light. Compatible light-colored stabilizers must be added when the black color is not acceptable. Clay and ground limestone have also been used as fillers. It is customary to characterize LDPE by its melt flow index, i.e., the weight in grams that is extruded in ten minutes at 190°C.

The principal end use of LDPE is as film and sheet. Polyethylene film is used as a mulch barrier, as a shrink protector for pallets, and as writing paper. About 15% of this polymer is injection molded and larger quantities are extruded as tubing and as paper coatings. Squeeze bottles are produced by blow molding LDPE. This polymer is also used as a coating for wire and cable.

A typical molded LDPE part will have the following properties: Tensile strength 1,500 psi, elongation 700%, impact strength, no break, Rockwell hardness D45, coefficient of expansion 15×10^{-5} in./in./°C, heat deflection temperature at 66 psi, 100°F, excellent electrical properties, good resistance to cold solvents, and good resistance to acids.

8.2 HIGH DENSITY POLYETHYLENE

The classical polyethylene produced in the laboratory by Von Peckmann in the last half of the nineteenth century was actually a linear polymer. Since a linear polymer chain occupies less volume than a branched polymer chain, it has a higher density and its structure corresponds to the product called high density polyethylene (HDPE). As mentioned in the preceding section, the product produced by the Wurtz reaction and investigated by Dr. Carothers was also a low molecular weight linear polymer.

The first commercially-feasible process for the production of HDPE was developed by Ziegler in Germany in 1954. A comparable product was also developed in the Phillips laboratories in 1955 and in the laboratories of Standard Oil Company of Indiana at a later date. The original name of low pressure polyethylene was used to describe this linear polymer, but the term high-density polyethylene is preferred. Over one million tons of this important plastic were produced in the USA in 1973.

According to ASTM standards, HDPE has a density of 0.941–0.965 g/cm³. Any polyethylene having a density of 0.926–0.940 g/cm³ is called *medium-density polyethylene*. The degree of crystallinity in the high density product is greater than 85% for the Ziegler product and about 95% for the

Phillips product. The Phillips process is favored for the production of HDPE in the USA but the Ziegler process is widely used in other countries.

The catalyst used by Ziegler to produce his first linear polyethylene was aluminumtriethyl, but the present practice is to use a mixture of titanium tetrachloride ($TiCl_4$) and aluminumtriethyl in an inert atmosphere. The actual catalyst is produced from a reaction of these reactants, but the complex chemistry involved is beyond the scope of this textbook. The polymerization is conducted at about 70°C at a pressure of about 80 psi using low-boiling alkanes as the solvent. It is customary to quench the reaction with ethanol and to remove the catalyst residues by solvent washing.

The Phillips catalyst system consists of chromic oxide (CrO_3) supported on silica and alumina. This catalyst is first activated by heating at 250°C and it is then suspended in cyclohexane. Ethylene is polymerized in this slurry at 100°C and at a pressure of 100–500 psi. Modifiers such as isobutane may also be added to control the reaction. The final product is dissolved and filtered to remove catalyst residues.

Metallic oxides, such as nickel oxide (Ni_2O_3) or molybdenum oxide (Mo_2O_3), on porous supports, such as charcoal or alumina, are used as catalysts in the Standard of Indiana process. These catalysts are activated by metallic hydrides such as sodium hydride (NaH) or calcium hydride (CaH_2), and the polymerization takes place at 230–270°C at a pressure of 40–80 atm.

HDPE is extruded to form corrugated drain pipe, blow molded to produce containers ranging in size from 1 gal milk containers to 55-gal drums, injection molded to produce 5 gal shipping pails, and extruded to produce filaments. Other HDPE end products are used as pipe, toys, closures, housewares, and fuel tanks.

The properties of a typical molded HDPE part are as follows: Tensile strength 3,500 psi, elongation 600%, impact strength 15 ft lbs/in. if notch, shore hardness D65, coefficient of expansion 112×10^{-4} in./in./°C, heat-deflection temperature at 264 psi 120°C, excellent electrical properties, good resistance to cold solvents, and good resistance to acids.

Polymers of ethylene with average molecular weights in the range of 1.5–3 million are sometimes called *ultra-high molecular weight PE* (UHMWPE). Since these polymers have a melt index of zero, they are more difficult to process. They are characterized by good ductility, good properties at low temperatures, and excellent resistance to abrasion.

The resistance of polyethylene coatings to elevated temperatures may be improved by crosslinking the polymer after applications. Exposure of polyethylene to high energy radiation, such as that produced by cobalt-60, will cause crosslinking. Comparable effects may be obtained when peroxy compounds such as dicumyl peroxide are added to the polymer. Such compositions may be molded or extruded and crosslinked by heating at a higher temperature.

8.3 POLYPROPYLENE

Polymers of propylene were prepared in the nineteenth century by adding Friedel-Crafts catalysts such as aluminum bromide to propylene at very low temperatures. However, these sticky low molecular weight amorphous products had no commercial value. Useful products were produced in 1954 by Natta who used a Ziegler-type catalyst system.

By polymerizing propylene at 70–80°C at 100 atmospheres pressure using a slurry of titanium trichloride ($TiCl_3$) and aluminumdiethyl chloride ($ClAl(C_2H_5)_2$), Natta obtained a highly-crystalline, isotactic polypropylene (PP). As stated previously in Sec. 1.5 and as shown in Figs. 1-2 and 8-2, the methyl groups in isotactic polypropylene are all on the same side relative to the polymer chain.

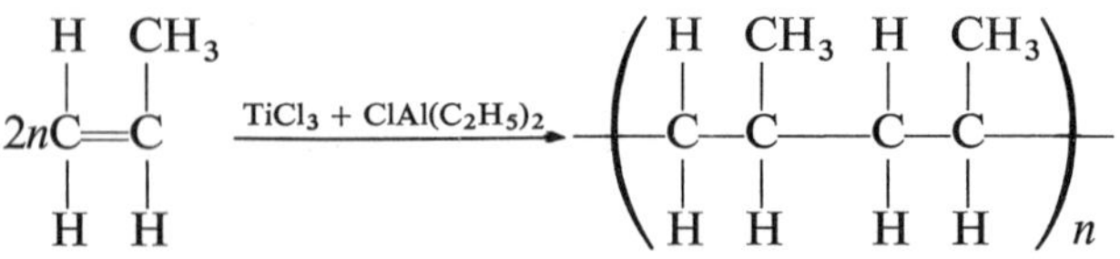

$$2n\underset{\underset{H}{|}}{\overset{\overset{H}{|}}{C}}=\underset{\underset{H}{|}}{\overset{\overset{CH_3}{|}}{C}} \xrightarrow{TiCl_3 + ClAl(C_2H_5)_2} \left(\underset{\underset{H}{|}}{\overset{\overset{H}{|}}{C}}-\underset{\underset{H}{|}}{\overset{\overset{CH_3}{|}}{C}}-\underset{\underset{H}{|}}{\overset{\overset{H}{|}}{C}}-\underset{\underset{H}{|}}{\overset{\overset{CH_3}{|}}{C}} \right)_n$$

Propylene   Polypropylene

Fig. 8-2 Equation for the polymerization of propylene.

Polypropylene is now produced in the USA at an annual rate of almost one million tons. The commercial product is obtained by removing a slurry of polymer and Ziegler-Natta catalyst in unreacted propylene after 35–70% of the monomer has been converted to the polymer. Catalyst residues are removed by washing the centrifuged product with methanolic hydrochloric acid, filtering, washing, and steam distilling. Any undesirable rubbery atactic polymer produced is soluble in hexane, and this solvent is used as a test to ascertain the extent of tacticity. The commercial product has an isotactic polymer-constant of more than 90% and a degree of crystallinity greater than 60%.

Polypropylene has a higher melting point and higher melt viscosity than HDPE. Hence the melt flow index is determined at 230°C. The increase in volume caused by the presence of methyl groups on alternate carbon atoms yields a product with the low density of 0.90 g/cm^3. Because of the presence of stereoregular bulky methyl groups, polypropylene retains its stiffness at temperatures up to 140°C.

As evident from its structural formula, shown in Fig. 8-2, the carbon atoms with the methyl groups have only one hydrogen atom. Since the latter is located on a tertiary carbon atom, it is more readily removed under oxidizing conditions than a hydrogen atom in HDPE. Thus, polymers like polypropylene readily form free radicals that may undergo degradation. Such oxidative degradation is retarded by the addition of about 0.5% each of a substituted phenolic antioxidant and an ester of a dithiocarboxylic acid, such as dilauryl dithiopropionate. Polymeric phenyl phosphites may also be used in place of the latter stabilizer. It is also customary to add substituted benzophenones or other ultraviolet light stabilizers to polypropylene.

Polypropylene is much more resistant to stress cracking than LDPE. It also has the ability to withstand cyclic bending and thus is used for molded hinges. Its excellent resistance to elevated temperatures can be improved by the addition of fillers such as talc, asbestos, or fibrous glass. Polypropylene becomes brittle at low temperatures.

Approximately one half of all polypropylene produced is injection-molded to form automotive parts, radio or television cabinets, and kitchenware. About one third of the total production is extruded as filaments used for brushes, ropes, carpets, and blankets. Fibers are also produced by the fibrillation of split films. The latter are used as floor mats and as synthetic grass turf.

About one sixth of the total production of polypropylene is used as film. Because of its high crystallinity, this film is translucent, but it can be made clearer than cellophane by orienting the crystals in the plane of the film by biaxially drawing the film. Microporous film can be produced by controlled crystallization. Pigmented polypropylene film is also used as a substitute for writing paper.

Pipe is produced by the extrusion of polypropylene. Large containers are made by molding or blow molding this polymer. Polypropylene sheets are thermoformed for the manufacture of luggage.

A typical-molded unfilled polypropylene part will have the following properties: Tensile strength 5,000 psi, flexural strength 7,000 psi, compressive strength 6,500 psi, impact strength 1.0 ft lbs/in. of notch, Rockwell hardness R95, coefficient of expansion of 8×10^{-5} in./in./°C, heat deflection temperature at 264 psi, 130°F, excellent electrical properties, and good resistance to solvents and acids.

8.4 OLEFIN COPOLYMERS

The homopolymer of propylene becomes brittle at low temperatures. This deficiency may be overcome by blending natural rubber, butyl rubber, polyisobutylene or other elastomers with polypropylene. Plastics that are satisfactory for low-temperature service may also be produced by copolymerizing propylene with ethylene.

As stated in Sec. 1.8, both ethylene and propylene mers are present in the backbone of ethylene-propylene copolymers. The correct name for this macromolecule is poly(ethylene-co-propylene). These products are obtained by the copolymerization of these monomers in heptane in the presence of Ziegler-Natta catalysts. There is little tendency for crystallization when the mer units are randomly distributed in the polymer backbone. Random ethylene-propylene (EPM) copolymers are flexible and are used commercially as elastomers.

In contrast, stereospecific propylene-ethylene copolymers, in which ethylene is the minor constituent, are useful plastics. However, as might be anticipated from a knowledge of polymer structure, the degree of crystallinity of these copolymers is less than that of the homopolymer. The flexibility, resistance to impact, and resistance to embrittlement at low temperatures of these copolymers are greater than that of polypropylene. The effect of copolymerization is dramatically demonstrated by the improved flexibility of the stereoblock copolymer of propylene. Block copolymers consist of long sequences of specific mer units in the polymer backbone. As shown in Fig. 8-3, the only difference between the structure of polypropylene and its stereoblock copolymer is the arrangement of the two different sequences of propylene units.

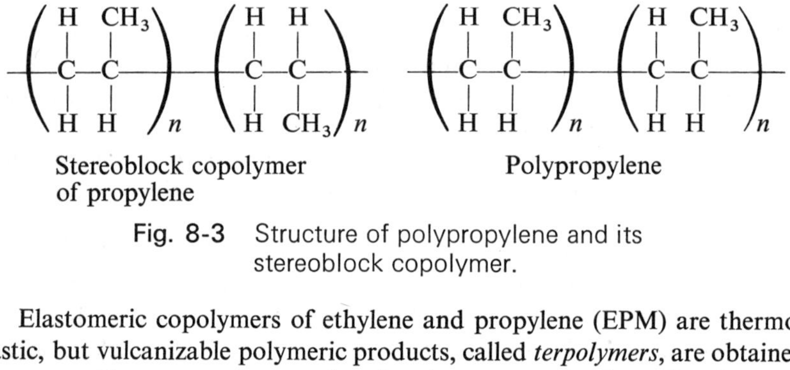

$$\left(\begin{array}{cc} H & CH_3 \\ | & | \\ C-C \\ | & | \\ H & H \end{array}\right)_n \left(\begin{array}{cc} H & H \\ | & | \\ C-C \\ | & | \\ H & CH_3 \end{array}\right)_n \quad \left(\begin{array}{cc} H & CH_3 \\ | & | \\ C-C \\ | & | \\ H & H \end{array}\right)_n \left(\begin{array}{cc} H & CH_3 \\ | & | \\ C-C \\ | & | \\ H & H \end{array}\right)_n$$

Stereoblock copolymer of propylene   Polypropylene

Fig. 8-3   Structure of polypropylene and its stereoblock copolymer.

Elastomeric copolymers of ethylene and propylene (EPM) are thermoplastic, but vulcanizable polymeric products, called *terpolymers*, are obtained when a small amount of a tetrafunctional monomer, such as dicyclopentadiene, is copolymerized with ethylene and propylene. These elastomers (EPDM) have a low specific gravity and are resistant to ozone.

A typical random copolymer of ethylene and propylene (EPM) will have the following properties: Density, 0.86 g/cm^3, tensile strength 500 psi, excellent electrical properties, and good resistance to solvents and acids. The tensile strength and hardness may be improved by reinforcing with fibrous glass. The resistance to creep of the vulcanized terpolymer (EPDM) is greater than that of the linear copolymer.

The block copolymers of propylene and a small amount of ethylene are prepared by a programmed rate of addition of propylene and ethylene to the reactor. The properties of these crystalline copolymers may be varied by controlling the monomer ratios in the feed. These macromolecules, called *polyallomers*, have good resistance to stress cracking and to low temperature embrittlement.

A typical polyallomer will have the following properties: Density 0.89 g/cm^3, tensile strength 3,500 psi, elongation 450%, impact resistance 4 ft lbs/in. of notch, Rockwell hardness R70, coefficient of expansion 9 × 10^{-5} in./in./°C, heat deflection temperature at 264 psi, 130°F, excellent electrical properties, and good resistance to solvents and acids.

Copolymers of ethylene and vinyl acetate were patented by Squires in 1946. These monomers are copolymerized under conditions similar to those used for the production of LDPE. Copolymers with a high vinyl acetate content are rubbery, but an ethylene (70)-vinylacetate (30) copolymer is a waxy product that is used in hot-melt compositions.

A product produced by the copolymerization of ethylene (93) and vinyl acetate (7) is softer and more flexible than linear polyethylene. This copolymer (EVA) is also more compatible with fillers. The properties of a typical molded EVA copolymer part are as follows: Tensile strength 2,000 psi, elongation 750 percent heat deflection temperature at 264 psi 93°F, good electrical properties, and fair resistance to solvents and acids. The properties of copolymers of ethylene and ethyl acrylate (EEA) are similar to those described for EVA having similar ethylene content.

Copolymers of ethylene and 1-butene or 1-hexene are produced by the Phillips process and have been available commercially since 1958. These copolymers, as those described previously, are resistant to environmental stress cracking and are used for blow molding of detergent bottles. Relatively clear plastic bottles are also blow molded from copolymers of propylene and vinyl chloride $\left(H_2C=\overset{H}{\underset{}{C}}Cl\right)$.

The term *ionomer* is used to describe copolymers of ethylene and small amounts of monomers containing carboxyl groups such as methacrylic acid ($H_2C=C(CH_3)COOH$). These products are produced by the techniques described previously for the production of LDPE. They have been available commercially since 1964. It is customary to form salts of these copolymers with sodium, potassium, magnesium, or zinc ions. As shown in Fig. 1-12, these saltlike products behave like crosslinked polymers at ordinary temperatures but may be injection molded and extruded at elevated temperatures.

These polyelectrolytes are transparent and retain their good physical properties at low temperatures. They are being used for packaging, extruded films and coatings, shoe soles, and golfball covers. A typical molded ionomer part will have the following properties: Density 0.95 g/cm^3, tensile strength 4,500 psi, elongation 400%, shore hardness D60, coefficient of thermal expansion 1.2 × 10^{-4} in./in./°C, heat deflection temperature at 264 psi, 190°F, poor electrical properties, excellent resistance to oils, greases, and other solvents, and fair resistance to acids.

8.5 OTHER POLYOLEFINS

Natural rubber (*Hevea braziliensis*) or polyisoprene is a polydiolefin. As stated in Sec. 1.10, natural rubber is a sticky product because of the unrestricted slippage of its polymer chains. Figure 8-4 shows that it is a cis isomer since the chain segments attached to the ethylenic groups $\left(-\underset{\text{H}}{\text{C}}=\underset{\text{H}}{\text{C}}-\right)$ are on the same side of the plane. In contrast, gutta percha, which is also a naturally occurring polymer, is a hard thermoplastic in which the chain segments attached to the ethylene groups are on the opposite sides of the plane:

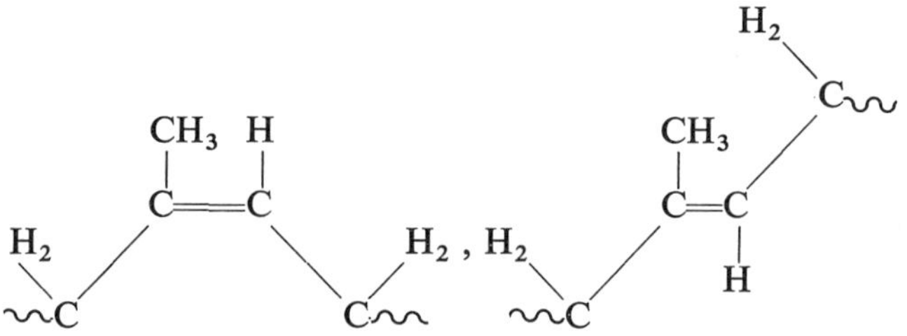

Segment of natural rubber (cis)   Segment of gutta percha (trans)

Fig. 8-4 Chain segments showing cis and trans structures in polyisoprenes.

Chain slippage is reduced by the symmetrical trans structure present in gutta percha. This naturally-occurring thermoplastic has been used as a cable coating and for golfball covers. Synthetic polydiolefins are available in a range of molecular weights ranging from liquids to solids. These polybutadienes may be compounded with sulfur or peroxy compounds and cured to produce thermoset compositions.

Homopolymers of isobutylene ($H_2C=C(CH_3)_2$) were commercially available in Germany prior to World War II under the trade name of Oppanol B. As shown earlier in Figs. 2-1, 2-2 and 2-3, the polymerization of isobutylene is initiated by a Lewis acid such as aluminum chloride in the presence of a trace of moisture (H_2O). A sticky, high molecular-weight product is produced when isobutylene is polymerized at $-80°$ to $-100°C$. These products may be blended with other more rigid polymers to produce flexible products with good adhesive characteristics.

The lower molecular-weight products are obtained at higher polymerization temperatures. They are used as additives to control the viscosity of lubricating oils. Because of its tendency to flow at ordinary temperatures, polyisobutylene has limited utility. This objection was overcome in the 1930s

Sec. 8.5   OTHER POLYOLEFINS

by Sparks and Thomas who copolymerized isobutylene with a small amount of isoprene to produce a vulcanizable elastomer called *butyl rubber*. A section of a butyl rubber polymer chain is depicted in Fig. 8-5.

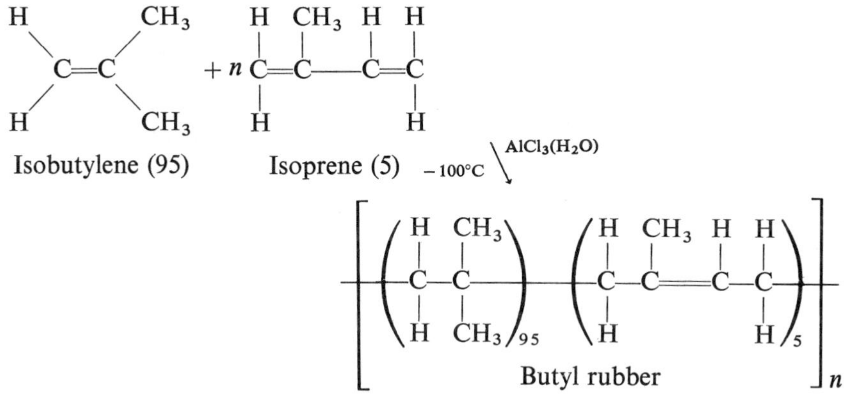

$$\underset{\text{Isobutylene (95)}}{\overset{H}{\underset{H}{>}}C=C\overset{CH_3}{\underset{CH_3}{<}}} + n\ \underset{\text{Isoprene (5)}}{\overset{H}{\underset{H}{|}}\overset{CH_3}{\underset{|}{C}}=C-C=C\overset{H}{\underset{H}{|}}} \xrightarrow[-100°C]{AlCl_3(H_2O)}$$

$$\left[\left(\begin{array}{cc}H & CH_3\\ | & |\\ -C-C-\\ | & |\\ H & CH_3\end{array}\right)_{95}\left(\begin{array}{cccc}H & CH_3 & H & H\\ | & | & | & |\\ -C-C=C-C-\\ | & & & |\\ H & & & H\end{array}\right)_{5}\right]_n$$

Butyl rubber

Fig. 8-5 Equation for the production of butyl rubber.

Because of the dearth of unsaturated groups in butyl rubber, this elastomer is more resistant than natural rubber to heat and ozone. Butyl rubber is extruded to form inner tubes, steam hose linings, cable coatings, and caulking compositions. Low molecular-weight polymers are also produced by the Friedel-Crafts ($AlCl_3$) polymerization of 1-butene.

In contrast, a high molecular-weight isotactic crystalline product is obtained by the polymerization of 1-butene in the presence of a Ziegler-Natta catalyst. This polymer has a specific gravity of 0.90, a heat resistance comparable to HDPE, and excellent electrical properties.

Poly(4-methylpentene-1) has a lower specific gravity (0.83) and a higher melting point (464°F) than poly(1-butene). It has been commercially available under the tradename TPX since 1965. This polymer is produced in Great Britain by the Ziegler-Natta polymerization of 4-methylpentene-1. It is customary to copolymerize this monomer with traces of other olefins in order to assure a high degree of transparency. It is used for electrical applications, packaging, injection-molded parts, laboratory ware, and as extruded film and coatings.

A typical molded poly(4-methylpentene-1) part would have the following properties: Density 0.83 g/cm^3, tensile strength 3,500 psi, elongation 20%, impact strength 1.0 ft lbs/in. of notch, Rockwell hardness L70, coefficient of expansion 1.2×10^{-4} in./in./°C, excellent electrical properties, and good resistance to solvents and acids.

8.6 MISCELLANEOUS POLYOLEFIN PLASTICS

Charles Goodyear produced a new composition of matter when he heated natural rubber with a small amount of sulfur over a century ago. Subsequently, his brother Nelson heated natural rubber with larger quantities of sulfur and produced a thermosetting plastic that he called *ebonite* or *hard rubber*. Similar reactions will take place when other unsaturated polymers are heated with sulfur.

In the early part of the 20th century, Harries and Kirchof changed the structure of the rubber molecule from a cis linear chain polymer to a cyclic polymer by heating it with a Friedel-Crafts catalyst, such as aluminum chloride. *Cyclized rubber*, which is a thermoplastic, was produced commercially in 1927. Solutions of these Pliolite resins have been used as adhesives and coatings, but less expensive polymers have replaced cyclized rubber in many applications.

The reaction of chlorine and natural rubber was investigated by Traun in 1859. A *chlorinated rubber* containing about 66% chlorine is available commercially under the trade name of Parlon. It is used as a coating for concrete surfaces. Staudinger investigated the reaction of a solution of rubber and gaseous hydrogen chloride in the 1920s. Subsequently, a *rubber hydrochloride* was introduced commercially under the trade name of Pliofilm. This product is used as a packaging film. Epoxidized polybutadienes are produced by the oxidation of a solution of polybutadiene by peracetic acid. These products may be cured by polyamines like epoxy resins.

Butyl rubber and polyethylene may be reacted with chlorine. *Chlorinated polyethylene* (Tyrin) is used as a modifier for other polymers. A product called Hypalon is produced by the chlorosulfonation of polyethylene. Copolymers of ethylene and propylene have also been chlorinated.

Water soluble polymers have been produced by the polymerization of ethylene oxide in the presence of strontium carbonate. These products may be molded or extruded and then reacted with a polycarboxylic acid to reduce the effect of moisture. When small amounts of *poly(ethylene oxide)* are added to water, the resistance to flow of the water is reduced. Thus high velocity water streams for fighting fires may be obtained by the addition of this polymer to the water used in the hoses. These aqueous streams may be pumped greater distances without increasing the water pressure.

BEHAVIORAL OBJECTIVES

The most widely used thermoplastics are polyolefins such as polyethylene and polypropylene. These are injection molded, extruded, and blow molded to produce a wide variety of items used in our daily lives.

GLOSSARY

After reading this chapter, you should understand the following concepts:

1. High density polyethylene is a linear polymer which is produced by the polymerization of ethylene in the presence of a Ziegler-Natta catalyst.
2. Low density polyethylene is a highly branched polymer which is produced by the free radical chain polymerization of ethylene at high pressure.
3. Polypropylene which has a lower density than HDPE is produced by the polymerization of propylene in the presence of a Ziegler-Natta catalyst. The commercial produce is isotactic PP.
4. Random copolymers of ethylene and propylene are flexible elastomers.
5. Copolymers of ethylene with vinyl acetate or ethyl acrylate are more flexible than the homopolymers of ethylene.
6. The salts of copolymers of ethylene and methacrylic acid which are readily molded behave like crosslinked polymers when cooled to room temperature.
7. Natural rubber is a flexible cis-polyisoprene, whereas gutta percha is a rigid trans-isomer of polyisoprene.
8. Polyisobutylene is a sticky polymer which is produced by the low temperature polymerization of isobutylene in the presence of a Lewis acid such as aluminum chloride.
9. Butyl rubber is a copolymer of isobutylene with a small amount of isoprene. Hence it may be cured or vulcanized like natural rubber.
10. Unsaturated polymers such as Hevea rubber may be cyclized, chlorinated, or hydrochlorinated to produce useful plastics.
11. Polymers of ethylene oxide which are water soluble decrease the resistance to flow when added to water.

GLOSSARY

Butyl rubber: a copolymer of isobutylene (95) and isoprene (5).
Cis: an isomer having functional groups on the same side of ethylenic carbon atoms.
Copolymer: a macromolecule consisting of more than one building block or mer in the polymer chain.
Cyclized rubber: the reaction product of natural rubber and aluminum chloride.
d: belonging to the d optical isomer family.
Ebonite: hard rubber.
EEA: poly(ethylene-co-ethyl acrylate).
Elastomer: a rubber like polymer.
EPM: poly(ethylene-co-propylene).
EVM: poly(ethylene-co-vinyl acetate).

Goodyear, Charles: discoverer of the process in which natural rubber is crosslinked or vulcanized by heating with sulfur.
Gutta percha: a stiff isomer of natural rubber.
HDPE: high density polyethylene.
Hevea rubber: natural rubber.
Homopolymer: a homogeneous polymer such as HDPE.
Ionomer: copolymers containing carboxylic groups.
Isotactic: a polymer with pendant groups arranged systematically on the same side of the polymer.
l: belonging to the l optical isomer family, i.e. a mirror image of the d isomer.
LDPE: low density polyethylene.
Lewis acid: an electron deficient molecule such as aluminum chloride.
Melt flow index: a measure of melt viscosity.
PE: polyethylene.
Phillips catalyst: chromic oxide supported on silica and alumina.
Polyallomer: an ethylene-propylene copolymer.
PP: polypropylene.
Stereo block copolymers: block polymers consisting of sequences of isotactic chains with different arrangements in space such as d,l.
Stereospecific: a polymer in which the building units have a definite arrangement in space.
Trans: an isomer having functional groups on alternating or opposite sides of the ethylenic carbon atoms.
TPX: poly(4-methylpentene-1).
UHMWP: ultrahigh molecular weight polyethylene.
Ziegler-Natta catalyst: a mixture of titanium trichloride and triethylaluminum.

QUESTIONS

1. Which has the greater volume, HDPE or LDPE?
2. Which is more highly branched, HDPE or LDPE?
3. Which is more resistant to elevated temperature, HDPE or PP?
4. Which is more rigid, isotactic PP or atactic PP?
5. Which is more flexible, HDPE or a copolymer of ethylene and vinyl acetate?
6. What is an ionomer?
7. Which will be more flexible, the cis or the trans isomer of polyisoprene?
8. How would you reduce the cold flow of polyiosbutylene?
9. Why aren't the plastics made by the cyclization, chlorination and hydrochlorination of rubber used to a greater extent?
10. How could you increase the rate of flow of water?

ANSWERS

1. LDPE.
2. LDPE.
3. PP.
4. Isotactic PP.
5. The copolymer of ethylene and vinyl acetate.
6. A copolymer of ethylene and methacrylic acid.
7. The cis isomer, i.e., natural rubber.
8. Copolymerize isobutylene with isoprene and vulcanize the copolymer of these two monomers.
9. Because the high costs of rubber and the subsequent reaction, the products are not economically competitive with comparable synthetic plastics.
10. Add a small amount of poly(ethylene oxide) to water to reduce its resistance to flow.

chapter 9

Polystyrene and Related Polymers

9.1 POLYSTYRENE

Styrene monomer was originally obtained by heating storax gum. Bonastre isolated this monomer $\left(\bigcirc\!\!-\!\!\underset{\underset{CH_2}{\overset{H}{|}}}{C}\right)$ that which was called *styrol*, in 1831 and in 1839, Simon studied the solid product that formed on standing. The original polymer was called *metastyrene* by Hoffman and Blyth who also investigated this macromolecule in 1845.

Considerable research was undertaken by Ostromislensky in the early 1900s. Polystyrene was patented in Great Britain by Kronstern in 1900. The name polystyrene was proposed by Staudinger in the 1930s. The first commercial polymer was produced in Germany in 1927. Polystyrene was produced commercially by Dow in the USA in 1937.

Sec. 9.1   POLYSTYRENE

Less than 200 thousand pounds of polystyrene were produced in 1938. This annual rate of production increased to almost 50 million pounds in 1945. Over four billion pounds of polystyrene and related plastics were produced in the USA in 1974. The production of polyolefins and vinyl plastics is now greater than that of polystyrene. However, the availability of inexpensive styrene monomer resulting from the World War II synthetic rubber crash program was largely responsible for starting the modern plastics industry. Future growth of polystyrene will be dependent on the cost and availability of benzene.

The original styrene from storax was probably obtained by the thermal decarbonation or removal of carbon dioxide from cinnamic acid. Berthelot produced styrene by the dehydrogenation or removal of hydrogen from ethylbenzene in 1869. It is of interest to note that the present commercial-production method is similar to that used over a century ago. Equations for both the classical and commercial synthesis of styrene are shown in Fig. 9-1.

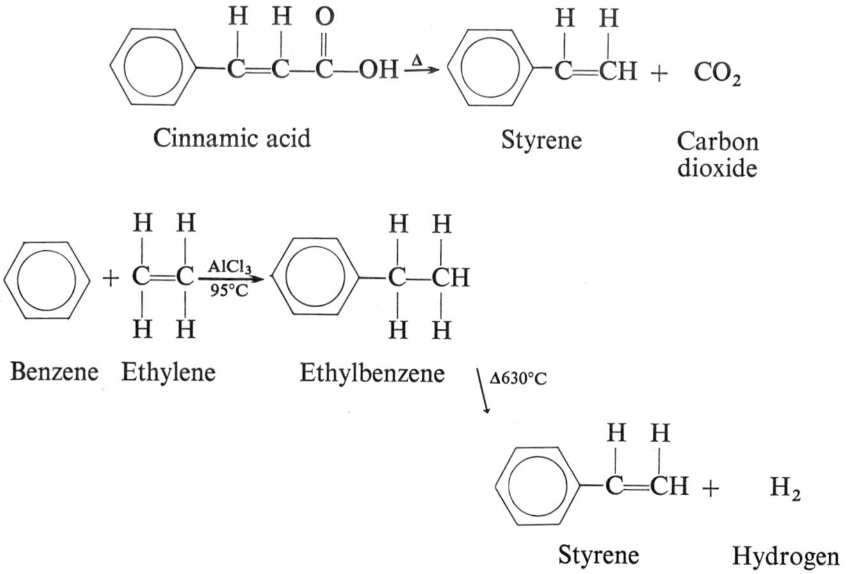

Ph–CH=CH–COOH $\xrightarrow{\Delta}$ Ph–CH=CH$_2$ + CO_2

Cinnamic acid → Styrene + Carbon dioxide

Ph–H + CH$_2$=CH$_2$ $\xrightarrow[95°C]{AlCl_3}$ Ph–CH$_2$–CH$_3$

Benzene + Ethylene → Ethylbenzene $\xrightarrow{\Delta 630°C}$

Ph–CH=CH$_2$ + H_2

Styrene + Hydrogen

Fig. 9-1 Equations for the synthesis of styrene.

Styrene is converted to polystyrene by cationic, free-radical or Ziegler-Natta catalyst polymerization techniques, such as those discussed in Sec. 2.2. It is customary to use a free radical initiator and to polymerize styrene by bulk or suspension techniques. Polystyrene can also be produced by solution or emulsion polymerization techniques. The general equation for the polymerization of styrene is depicted in Fig. 9-2.

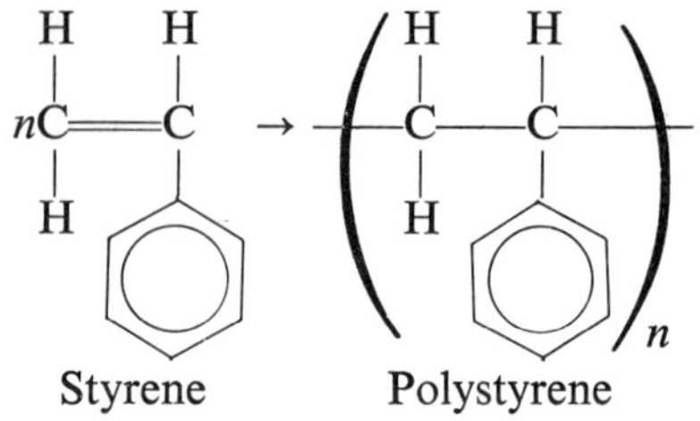

Fig. 9-2 Equation for the polymerization of styrene.

Commercial polystyrene (PS), sometimes called *crystal styrene*, is an actactic, amorphous (noncrystalline), clear, brittle polymer that tends to yellow when exposed to sunlight for long periods of time. Polystyrene is soluble in benzene, is softened by boiling water, and has excellent electrical and optical properties. The latter is related to its refractive index of 1.592. This high value provides brilliance to unpigmented grades of general purpose polystyrene and permits molded pieces to transmit light through curved sections.

Polystyrene may be injection molded or extruded. It is customary to add pigments or colorants and flame retardants in an extruder. Fibrous glass may also be added to increase the strength and resistance to elevated temperatures. As mentioned previously in Sec. 4.3, expanded polystyrene may be produced by the extrusion of polystyrene in the presence of a low-boiling liquid or by the molding of polystyrene beads that have been impregnated with pentane. Expanded polystyrene is used for insulation, buoys, and packaging.

The physical properties of polystyrene sheet are improved by biaxial orientation. Molded, general-purpose polystyrene is used for many applications where a brittle plastic is acceptable. For example, polystyrene is used for the production of wall coverings, housewares, cabinets, containers, and brittle toys. The properties of a typical molded polystyrene article are as follows: Specific gravity 1.05, tensile strength, 7,000 psi, flexural strength 10,000 psi, compressive strength 12,000 psi, impact strength 0.25 ft lbs/in. of notch, Rockwell hardness M65, coefficient of expansion 7×10^{-5} in./in./°C, heat deflection temperature 200°F, excellent electrical properties, fair resistance to oils and greases, and good resistance to acids.

9.2 POLYSTYRENE BLENDS

The production of brittle toys from general-purpose polystyrene has had an adverse effect on the creditability of all plastics. Fortunately, this deficiency can be overcome to some extent by blending the brittle plastic with a synthetic rubber, such as styrene-butadiene elastomer (SBR). These rubber modified polystyrene plastics may be produced by mixing the two polymers on a mill or in a heavy duty mixer.

The opaque composites obtained are somewhat less brittle than polystyrene. Latex blends have also been made, but the improvement in impact resistance of these coagulated composites over that of polystyrene are marginal.

In the preferred technique, the elastomer is dissolved in the styrene monomer before polymerization. The opaque polymerized product may contain up to 20% of rubber. It has an impact resistance that is superior to that of polystyrene. Presumably some of the styrene grafts onto the backbone of the elastomeric polymer. Thus, the elongation of a styrene 50-elastomer 50-blend produced by the preferred polymerization technique is three times greater than the polyblend produced by milling the polymers.

It has been suggested that the components of these blends or alloys are not entirely compatible. Thus, crack propagation is retarded by absorption of energy by the elastomeric particles in the composite. Superior high-impact grades of polystyrene have been produced by replacing the SBR by poly(cis-1,4-butadiene) or ethylene-propylene copolymers.

A typical part produced by molding a blend of polystyrene and rubber will have the following properties: Tensile strength 4,000 psi, flexural strength 7,500 psi, compressive strength 6,000 psi, impact resistance 3.0 ft lbs/in. notch, Rockwell hardness M50, coefficient of expansion 1.5×10^{-4} in./in./°C, heat deflection temperature at 264 psi, 190°F, very good electrical properties, fair resistance to oils and grease, and good resistance to acids.

As stated earlier, solutions of unsaturated polyesters in styrene monomer are polymerized to produce the so-called polyester plastics. Solutions of drying oils in styrene are also polymerized to produce paint resins. More compatible products are obtained when vinyltoluene is used in place of styrene.

9.3 COPOLYMERS OF STYRENE

While some improvement in properties is noted when two homopolymers are blended together, the improvement is much greater when the two monomers are polymerized together to produce copolymers. (See Sec. 1.8.) Styrene has been copolymerized with isobutylene by use of a Friedel-Crafts catalyst at low temperatures, but most of the commercial styrene copolymers are produced by use of peroxycompounds as initiators. The unsaturated polyester plastics, which were discussed in Chap. 8, are produced by the copolymerization of styrene and an unsaturated polyester in the presence of an initiator such as benzoyl peroxide.

Many of the pioneer styrene copolymers were produced in order to

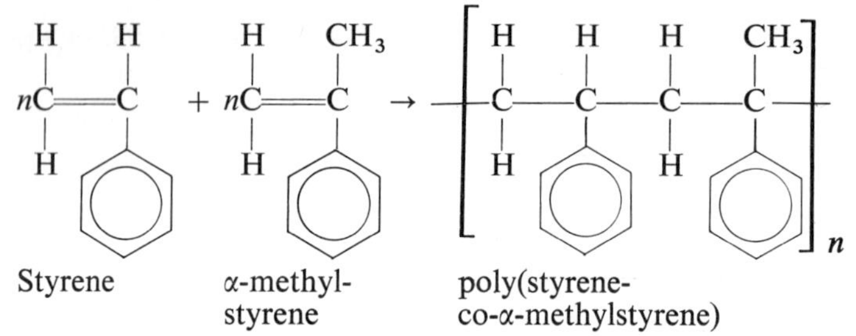

$$n\overset{H}{\underset{H}{C}}=\overset{H}{\underset{\phi}{C}} + n\overset{H}{\underset{H}{C}}=\overset{CH_3}{\underset{\phi}{C}} \rightarrow \left[\overset{H}{\underset{H}{C}}-\overset{H}{\underset{\phi}{C}}-\overset{H}{\underset{H}{C}}-\overset{CH_3}{\underset{\phi}{C}} \right]_n$$

Styrene   α-methyl-styrene   poly(styrene-co-α-methylstyrene)

Fig. 9-3 Equation for the formation of a styrene-α-methylstyrene copolymer.

obtain a more heat-resistant product. For example, the copolymer of α-methylstyrene and styrene, shown in Fig. 9-3, has a higher heat deflection temperature than polystyrene.

Other pioneer products with greater heat resistance than polystyrene were produced by the copolymerization of styrene with dichlorostyrene, fumarodinitrile, acrylonitrile, or p-divinylbenzene. The structural formulas for these monomers are given in Fig. 9-4.

Copolymers of styrene and divinylbenzene, in which the latter is the minor component, are insoluble, infusible, crosslinked macromolecules. Articles of these copolymers are produced by machining cast billets of the copolymer. The commercial copolymer of styrene (76) and acrylonitrile (24) is a transparant product that is also available as a fibrous glass-reinforced plastic. These plastics are used for applications where the heat deflection temperature requirements are greater than that of polystyrene.

A typical molded styrene-acrylonitrile copolymer will have the following

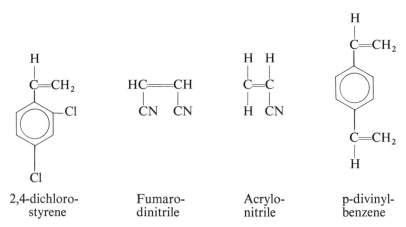

2,4-dichloro-styrene   Fumaro-dinitrile   Acrylo-nitrile   p-divinyl-benzene

Fig. 9-4 Structural formulas for monomers used to produce styrene copolymers.

properties: Tensile strength 10,500 psi, flexural strength 16,000 psi, compressive strength 16,000 psi, impact strength 0.45 ft lbs/in. of notch, coefficient of expansion 3.7×10^{-5} in./in./°C, heat deflection temperature at 264 psi, 205°F, good electrical properties, fair resistance to solvents, and good resistance to acids. This copolymer tends to yellow when used in outdoor service.

Copolymers of styrene with methyl methacrylate, ethyl acrylate, diethyl maleate and maleic anhydride have been produced commercially. Mixtures of homopolymers of styrene and methyl methacrylate are not compatible but copolymers such as poly(methyl methacrylate (40)-co-styrene) (60) are clear moldable plastics. The properties of these copolymers are in between those of the homopolymers and are related to the composition. They have higher heat-deflection temperatures and poorer resistance to sunlight than poly(methyl methacrylate).

The properties of a typical methyl methacrylate-styrene copolymer are as follows: Tensile strength 10,000 psi, flexural strength 17,500 psi, compressive strength 13,000 psi, impact strength 0.3 ft lbs/in. of notch, Rockwell hardness M75, coefficient of expansion 7×10^{-5} in./in./°C, heat deflection temperature at 264 psi 210°F, fair electrical properties, good resistance to mineral oil and ethanol, and good resistance to acids.

Comparable moldable products with similar superior heat resistance are produced by the copolymerization of alpha-methylstyrene and methyl methacrylate. These clear molded copolymers have been used for lenses, surgical items, and kitchen appliances.

As shown in Fig. 9-5, the styrene-maleic anhydride copolymer is hydrolyzed, i.e., converted to an acid when it is heated with an aqueous solution of sodium hydroxide. Thus, the water soluble product is widely used as a coating for textiles, as a drilling mud additive, and as an ingredient in latex paints and waxes.

The most widely used synthetic rubber is a copolymer of styrene (25) and butadiene (75), known by the symbol SBR. As shown in Sec. 9-2, blends of this copolymer and polystyrene have been used in applications that are not suitable for brittle plastics like polystyrene. Copolymers of styrene and butadiene, in which the former is the major component, have also been used when better impact strength was required.

Styrene-butadiene block copolymers, classified as thermoplastic rubbers, are also used when flexibility at low temperatures is essential. Presumably, intermolecular forces attract the polystyrene blocks and form large aggregates or domains at low temperatures. Thus, the flexible polybutadiene blocks are immobilized until the temperature is increased enough to destroy the polystyrene domains. The reduction of the intermolecular forces between the polystyrene blocks enhances the mobility of the polybutadiene blocks. Thus, these block copolymers are characterized by lower heat deflection temperatures than polystyrene.

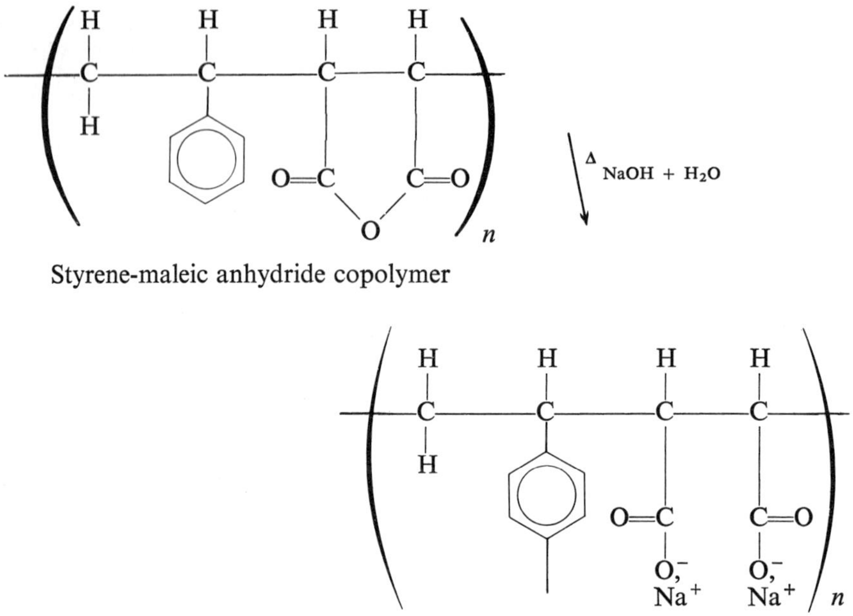

Styrene-maleic anhydride copolymer

Salt of Styrene-maleic acid copolymer

Fig. 9-5 Equation for the hydrolysis of styrene-maleic anhydride copolymer.

Styrene-butadiene block copolymers are produced under the trade name of Kraton. They are used for molded shoe soles, shoe uppers, molded flexible articles, sealants, coatings, and adhesives. Since they are soluble in acetone, ethyl acetate, and hydrocarbon solvents, they may be adhered to film or articles of this copolymer by these solvents using a so-called solvent welding technique.

9.4 ABS PLASTICS

Section 9-2 showed that the impact resistance of polystyrene has been improved by blending with elastomers. By the extension of this art, plastic engineers at Uniroyal Corporation were able to produce superior blends or alloys by simply coagulating mixtures of lattices of poly(styrene (70)-co-acrylonitrile (30)) and poly(butadiene (65)-co-acrylonitrile (35)). The original blends, patented in 1948, were produced under the trade names of Royalite and Kralastic. Cycolac is another trade name for ABS plastics.

Subsequently, production techniques were modified so that graft copolymers instead of mixtures were produced. The term ABS copolymer is now used to describe any of these so-called engineering plastics that are produced from various combinations of acrylonitrile (A), butadiene (B) and styrene (S). The annual production of ABS plastics in the USA is almost one billion pounds.

ABS copolymer plastics may be extruded as pipe and sheet, injection molded, and calendered. The sheets may be thermoformed and the molding powders may be molded in the presence of volatile liquids to produce cellular products. This versatile copolymer has also been blended with poly(vinyl chloride) (PVC) to produce an impact resistant product. The pioneer ABS plastics were opaque but transparent modifications are now available.

In ABS plastics, the styrene-acrylonitrile copolymer, acting as the continuous phase, provides high strength and some degree of rigidity at normal service temperatures. The butadiene-acrylonitrile elastomer, acting as the dispersed or discontinuous phase, provides toughness and improved resistance to impact. The graft copolymer, acting as a boundary layer between the continuous and dispersed phases, also provides a weak bonding force between these incompatible phases. Outstanding properties can be built into ABS plastics by proper selection of the composition and relative amounts of each of these components.

Self-extinguishing and fibrous glass reinforced modifications are commercially available. ABS plastics are used for automotive parts, telephones, conduit and pipe, household appliances, and furniture. The properties of a typical molded high impact ABS article are as follows: Tensile strength 6,000 psi, flexural strength 8,500 psi, compressive strength 6,000 psi, impact strength 6 ft lbs/in. of notch, Rockwell hardness R90, coefficient of expansion 1.1×10^{-4} in./in./°C, heat deflection temperature of annealed specimens at 264 psi, 210°F, fair electrical properties, poor resistance to organic solvents, and good resistance to acids.

9.5 POLYMERS OF STYRENE DERIVATIVES

Alpha-methylstyrene does not polymerize in the presence of free radicals but polymerizes readily in the presence of cationic initiators such as aluminum chloride at low temperatures such as $-70°C$. Indene and coumarone, obtained by the distillation of coal tar, are also polymerized by cationic initiators. These coal tar resins, which were obtained by Kraemer and Spilker in Germany during World War I, are usually produced by the addition of sulfuric acid to a solution of these so-called light-oil fractions. The structural formulas for these monomers are shown in Fig. 9-6.

Coumarone-indene resins are available in a wide range of softening temperatures under the trade names of Cumar, Nevindene, and Piccoumaron.

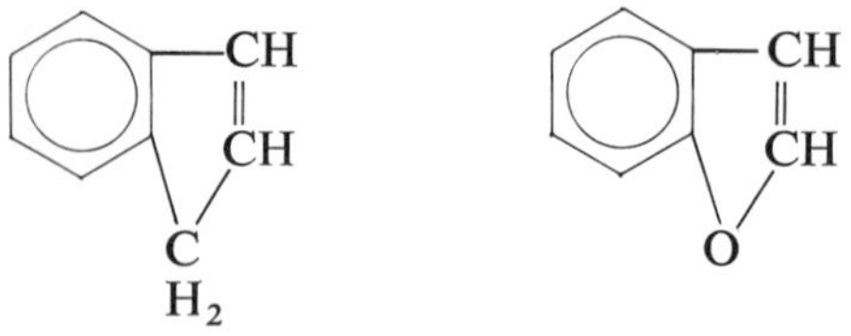

Indene Coumarone

Fig. 9-6 Structural formulas for indene and coumarone.

These resins may be hydrogenated to produce a more color-stable product called *Nevillite*. Comparable resins are also obtained from petroleum. These inexpensive resins are produced at an annual rate of over three hundred million pounds.

Polymers obtained by the cationic polymerization of α- and β-pinene are called *terpene resins*. They are lighter in color than the coumarone-indene resins and are used for similar applications. Figure 9-7 shows that these monomers are alicyclic and not aromatic. β-pinene contains a polymerizable methylene group (=CH_2). α-pinene is converted to polymerizable dipentene in the presence of cationic initiators such as sulfuric acid.

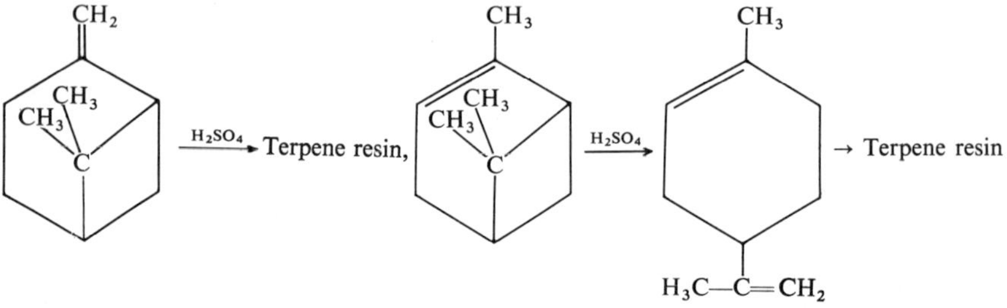

β-pinene α-pinene Dipentene

Fig. 9-7 Equations for conversion of α- and β-pinene to polymers.

Poly(-2-methylstyrene) or poly(2-vinyltoluene) has a higher heat deflection temperature than polystyrene. A mixture of meta and para-methylstyrene isomers is obtained by the dehydrogenation of the product obtained by the addition of ethylene to toluene. This mixture is added to unsaturated polyesters but is seldom polymerized in the absence of other monomers. Homopolymers of meta- and para-methylstyrene have lower heat-deflection temperatures than polystyrene.

Since chlorine groups on the benzene nucleus increase the intermolecular attraction of polychlorostyrenes, these polymers have higher heat deflection temperatures than polystyrene. Atactic poly(2,4-dichlorostyrene) has a heat deflection temperature of 120°C and a specific gravity of 1.40.

Cation exchange resins are produced by reacting crosslinked polystyrene with fuming sulfuric acid. Anion exchange resins have been produced by reducing nitrated crosslinked polystyrene or by reacting crosslinked poly(p-chloromethylstyrene) with ammonia. Since esters of amino acids will also react with this chloromethylated crosslinked polymer, the latter is used in the Merriefield synthesis for the synthesis of polypeptides.

9.6 AROMATIC HYDROCARBON POLYMERS

Stable at temperatures up to 300°C, polyxylylenes, have been produced since 1965 by the pyrolytic oxidative coupling of p-xylene $\left(H_3C-\bigcirc-CH_3\right)$. These polymers, called *parylenes*, are supplied in the form of a dimer $\left(-\overset{H_2}{\underset{|}{C}}-\bigcirc-\overset{H_2}{\underset{|}{C}}-\right)_2$ which is obtained by heating p-xylene at 950°C. In practice, the dimer is vaporized and pyrolyzed at 550°C to produce a diradical $\left(\cdot\overset{H_2}{\underset{|}{C}}-\bigcirc-\overset{H_2}{\underset{|}{C}}\cdot\right)$ that polymerizes in situ on any surface at temperatures below 70°C. Greater resistance to elevated temperatures may be obtained by using dimers of dichloro-para-xylene. The transparent films formed by polymerization in situ are used for coating circuit boards, heat exchangers, electronic components, and as membranes.

Another moldable engineering plastic was produced in 1966 by the oxidative coupling of 2,6-xylenol $\left(\bigcirc\begin{smallmatrix}-CH_3\\-OH\\-CH_3\end{smallmatrix}\right)$ by oxygen in the presence of copper salts. The repeating unit in the polymer chain is $\left(\bigcirc\begin{smallmatrix}-CH_3\\-O-\\-CH_3\end{smallmatrix}\right)$. Fibrous glass-reinforced and flame retardant grades of this polymer are available commercially. These thermoplastic resins are produced under the trade name Noryl and may be blended with polystyrene.

The *poly(aryl ethers)* are used for automotive parts, electrical appliances, and for molded parts that must withstand high temperatures. A typical molded poly(aryl ether) will have the following properties: Tensile strength 8,500 psi, flexural strength 13,000 psi, compressive strength 16,000 psi, impact strength 5 ft lbs/in. of notch, Rockwell hardness R117, coefficient of linear expansion 5×10^{-5} in./in./°C, heat deflection temperature at 264 psi, 230°F, excellent electrical properties, solubility in aromatic solvents, and good resistance to acids.

Poly(phenylene sulfides), produced under the trade name of Ryton, are also used for high-temperature service. Unlike the corresponding poly-(phenylene oxides), these plastics are also resistant to organic solvents. Poly(phenylene oxide) is used as a coating and as a moldable plastic. The properties of a typical part molded from this plastic are as follows: Tensile strength 10,500 psi, flexural strength 20,000 psi at break, impact strength 0.3 ft/lbs/in. of notch at 75°F and 1.0 ft lbs/in. of notch at 300°F, Rockwell hardness R124, deflection temperature at 264 psi, 278°F, coefficient of expansion 5.5×10^{-5} in./in./°C, good electrical properties, excellent resistance to solvents, and good resistance to acids.

The *poly(arylsulfones)*, produced under the trade name Astrel, also retain their physical properties at elevated temperatures. As shown in Fig. 9-8, these thermoplastics are produced by the condensation of the sodium salt of 2,2-bis-(4-hydroxyphenyl) propane (bisphenol A) and 4,4'-dichlorodiphenyl sulfone. The molecular weight is controlled by adding measured amounts of monofunctional organic chloride such as methyl chloride (CH_3Cl) to the reactants.

Since the sulfone group (O=S=O) is a chain-stiffening group, these

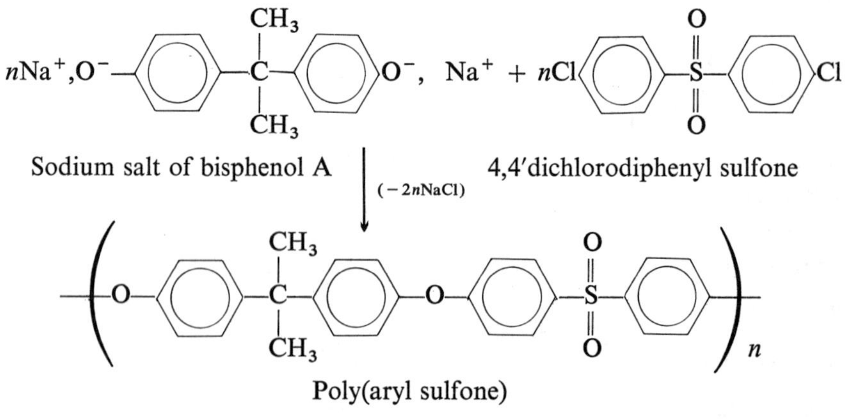

Fig. 9-8 Synthesis of poly(aryl sulfone).

Sec. 9.6 AROMATIC HYDROCARBON POLYMERS 189

amorphous plastics are extremely rigid. However, some flexibility in the backbone is provided for by the flexible ether and methylene groups that are linked to some of the phenyl groups in the chain. In spite of the presence of these flexible groups, the glass-transition temperature of these plastics is 525°F. In contrast, poly(alkyl sulfones) are characterized by low-glass transition temperatures and poor resistance to oxidation.

Poly(aryl sulfones) may be injection molded or extruded on standard processing equipment. These plastics are used for meter housings, light-fixture sockets, aircraft components, household appliances, hardware, and film. A typical molded part will have the following properties: Tensile strength 13,000 psi, flexural strength 17,000 psi, compressive strength 18,000 psi, impact strength 5.0 ft lbs/in. of notch, Rockwell hardness M110, coefficient of expansion 4.7×10^{-5} in./in./°C, heat deflection temperature at 264 psi 525°F, excellent electrical properties, excellent resistance to oil and grease but not to ketones and aromatic hydrocarbons, and excellent resistance to acids.

Another thermoplastic, called *phenoxy resin*, was introduced in the USA in 1962. Actually, as shown by the structural formula in Fig. 9-9, phenoxy resins are high molecular-weight, amorphous, linear pseudoepoxy resins that do not contain epoxy groups. Because of the presence of hydroxyl groups (OH), these transparent plastics may be cured or crosslinked by diisocyanates or anhydrides of dicarboxylic acids.

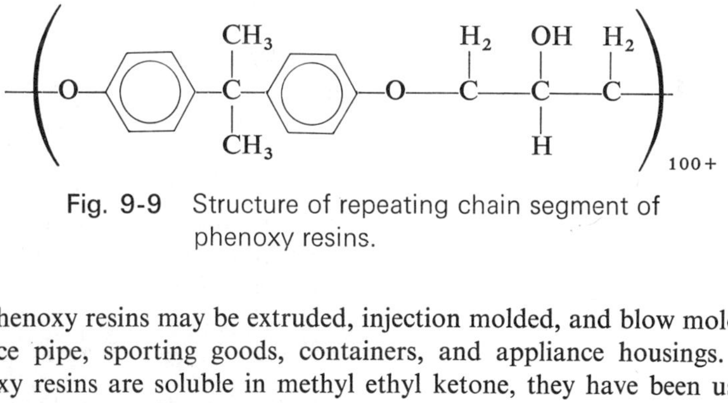

Fig. 9-9 Structure of repeating chain segment of phenoxy resins.

Phenoxy resins may be extruded, injection molded, and blow molded to produce pipe, sporting goods, containers, and appliance housings. Since phenoxy resins are soluble in methyl ethyl ketone, they have been used as adhesives and crosslinkable protective coatings. The properties of a typical injection molded phenoxy part are as follows: Tensile strength 9,000 psi, flexural strength 13,000 psi, compressive strength 11,000 psi, impact strength 5 ft lbs/in. of notch, Rockwell hardness R120, coefficient of expansion 6×10^{-5} in./in./°C, heat deflection temperature at 264 psi 180°F, good electrical properties, good resistance to mineral oils, and good resistance to acids.

BEHAVIORAL OBJECTIVES

Polystyrene, which was one of the original synthetic thermoplastics, is produced in large quantities by the free radical chain polymerization of styrene. After reading this chapter on this polymer and related polymers, you should understand the following:

1. Polystyrene has been produced commercially in the USA since 1937 by the polymerization of styrene monomer. The atactic brittle polymer is readily injection molded and extruded to produce a wide variety of products ranging from toys to dishware.
2. Blends of polystyrene and natural rubber are opaque, impact resistant composites. Mechanical working such as milling improves the properties of these blends.
3. Copolymers of styrene with acrylonitrile have better impact resistance and also are more heat resistant than PS.
4. Copolymers of styrene and maleic acid are water soluble.
5. Block copolymers of styrene and butadiene may be injection molded. They are called thermoplastic rubbers.
6. Macromolecules with grafts of polystyrene on a flexible backbone of a butadiene-acrylonitrile copolymer are widely used as engineering plastics.
7. Polymers may be produced from unsaturated hydrocarbons such as indene and dipentene by cationic polymerization with sulfuric acid at low temperature.
8. Ion exchange resins may be produced by the introduction of acidic or basic functional groups on the benzene rings in PS.
9. Polyxylenes are heat resistant thermoplastics which are obtained by the polymerization of a xylene diradical in situ.
10. Poly(aryl oxides) and sulfides are heat resistant engineering plastics obtained from the corresponding substituted phenols and thiophenols.
11. Poly(aryl sulfones) are also heat resistant engineering plastics obtained by condensation of sodium salts of dihydroxy aromatic hydrocarbons with dichloro aromatic sulfones.
12. Phenoxy resins are engineering thermoplastics having a structure that resembles epoxy resins.

GLOSSARY

ABS: copolymers of acrylonitrile, butadiene and styrene.
Amorphous: noncrystalline.
Biaxial orientation: stretching a plastic film in two directions.
Block copolymer: those with sequences of two or more building blocks in the same linear chain.
Cations: positively charged atoms or molecules.

Engineering plastics: those with good physical properties.
Graft copolymers: those with branches consisting of a sequence of other building blocks.
Mineral rubber: polyindene or polyterpenes.
PPO: a symbol used for polymers produced by the oxidative coupling of 2,6-xylene.
PS: polystyrene.
SBR: general purpose synthetic rubber, i.e., poly(butadiene-co-styrene).
Solvent welding: adhering by use of solvents.

QUESTIONS

1. What is the principal objection to the use of PS in toys?
2. Why is rubber blended with PS?
3. What is the principal objection to copolymers of styrene and acrylonitrile?
4. How can you make the copolymer of styrene and maleic anhydride soluble in water?
5. What is the advantage of a thermoplastic rubber over natural or synthetic rubber?
6. Why is ABS preferred over PS for extruded sheet, pipe, and molded parts?
7. Why must unsaturated hydrocarbons such as indene and dipentene be polymerized at low temperatures?
8. What cations may be removed from hard water by cation exchange resins?
9. Why is the production of polyxylene unique?
10. Why are poly(aryl oxides), poly(aryl sulfides), poly(aryl sulfones), and phenoxy resins of particular interest?

ANSWERS

1. Its inherent brittleness.
2. To improve the impact resistance of PS.
3. Yellow color.
4. Hydrolyze the anhydride groups with aqueous sodium hydroxide.
5. The block copolymer can be injection molded and does not require vulcanization.
6. Because of its superior resistance to impact.
7. In order to obtain high molecular weight products.
8. Calcium, magnesium, and iron ions.
9. Because it is produced by lowering rather than increasing the temperature of polymerization.
10. They are engineering plastics which may be used at relatively high temperatures.

chapter 10

Poly(Vinyl Chloride) and Related Polymers

10.1 RIGID POLY(VINYL CHLORIDE) (PVC)

Presumably, monomeric vinyl chloride $\left(\mathrm{H_2C}{=}\mathrm{\overset{H}{C}Cl}\right)$ was first produced by Liebig. Its preparation was described in 1835 by his student Regnault. The widely-used term *vinyl* was probably coined to describe this monomer in the 1840s. Vinyl is sometimes used to describe any unsaturated monomer.

The formation of the polymer when the monomer is exposed to sunlight is said to have been noted by Regnault, but this statement is controversial. Many believe that Regnault actually observed the polymerization of vinylidene chloride ($\mathrm{H_2C}{=}\mathrm{CCl_2}$) instead of vinyl chloride. It is also known that Baumann observed the conversion of gaseous vinyl chloride to a solid in the presence of sunlight in 1872.

That vinyl chloride could be polymerized in the presence of peroxides and ultraviolet light was noted in the first part of the 20th century by Klatte and Rollett, and Ostromislensky, respectively. It is of interest to note that the latter considered PVC polymers with different molecular weights to be different forms of the polymer.

Poly(vinyl chloride) was produced commercially in Germany and the USA in the late 1930s. Over four billion pounds of vinyl plastics were produced in the USA in 1974 and over 80% of this product was poly(vinyl chloride). Future growth will depend on industry's ability to solve the toxicological problems associated with the monomer (VCM).

Most of the vinyl chloride monomer is now produced by the thermal dehydrochlorination of dichloroethane $\left(\text{ClC}\overset{H_2}{-}\overset{H_2}{\text{CCl}}\right)$. The hydrogen chloride (HCl) byproduct is then oxidized and the chlorine obtained is added to the ethylene to produce more dichloroethane. Vinyl chloride is a gas with a boiling point of $-13°C$.

This monomer is polymerized by bulk, emulsion, and suspension techniques, but the latter is the preferred commercial process. Equations for the oxychlorination process and polymerization of vinyl chloride are shown in Fig. 10-1.

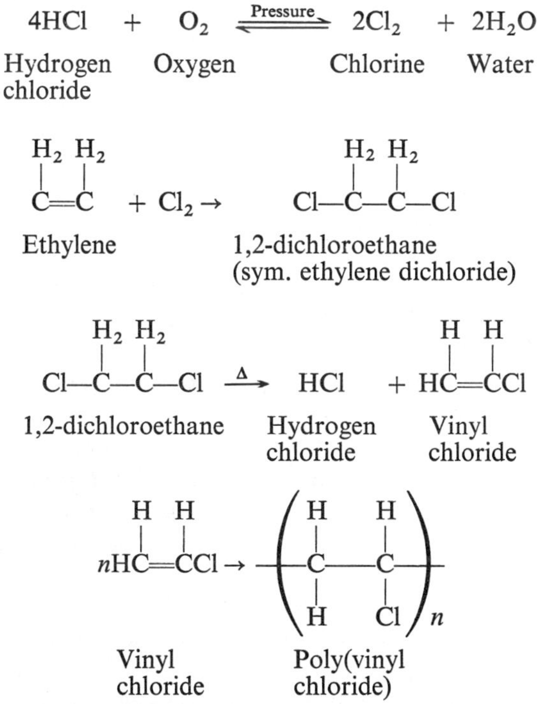

$$4HCl + O_2 \underset{}{\overset{\text{Pressure}}{\rightleftarrows}} 2Cl_2 + 2H_2O$$

Hydrogen chloride, Oxygen, Chlorine, Water

$$\underset{\text{Ethylene}}{H_2C=CH_2} + Cl_2 \rightarrow \underset{\substack{\text{1,2-dichloroethane} \\ \text{(sym. ethylene dichloride)}}}{Cl-CH_2-CH_2-Cl}$$

$$\underset{\text{1,2-dichloroethane}}{Cl-CH_2-CH_2-Cl} \overset{\Delta}{\rightarrow} \underset{\substack{\text{Hydrogen} \\ \text{chloride}}}{HCl} + \underset{\substack{\text{Vinyl} \\ \text{chloride}}}{HC=CHCl}$$

$$n\underset{\text{Vinyl chloride}}{HC=CHCl} \rightarrow \underset{\text{Poly(vinyl chloride)}}{\left(-\overset{H}{\underset{H}{C}}-\overset{H}{\underset{Cl}{C}}-\right)_n}$$

Fig. 10-1 Synthesis and polymerization of vinyl chloride.

Rigid poly(vinyl chloride) (PVC) is brittle and deteriorates at elevated temperatures, such as above 100°C or in the presence of sunlight. Figure 10-2 shows that this deterioration results from the dehydrochlorination which yields a polyunsaturated discolored product. The color of the originally-colorless plastic changes rapidly to yellow, orange, brown, and finally to black as the number of alternating or conjugated double bonds in the polymer increases.

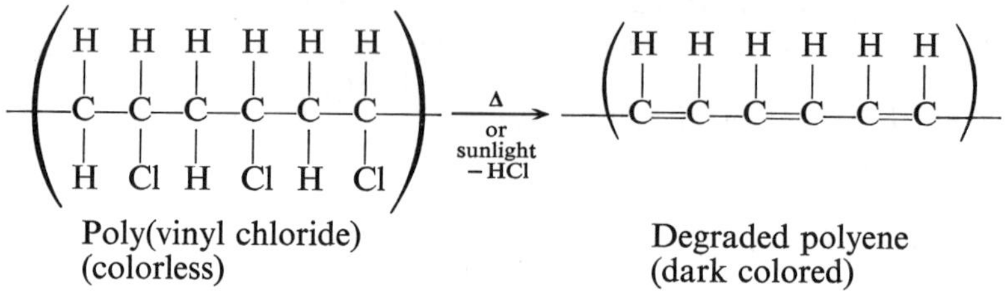

$$\left(\begin{array}{cccccc} H & H & H & H & H & H \\ | & | & | & | & | & | \\ -C & -C & -C & -C & -C & -C- \\ | & | & | & | & | & | \\ H & Cl & H & Cl & H & Cl \end{array} \right) \xrightarrow[\text{sunlight} \atop -HCl]{\Delta \text{ or}} \left(\begin{array}{cccccc} H & H & H & H & H & H \\ | & | & | & | & | & | \\ -C=C & -C=C & -C=C- \end{array} \right)$$

Poly(vinyl chloride) (colorless)   Degraded polyene (dark colored)

Fig. 10-2 Degradation of poly(vinyl chloride) by heat or sunlight.

As discussed in Secs. 3.1 and 3.2, this degradation and subsequent chain cleavage is retarded by the addition of appropriate stabilizers. Presumably, degradation takes place by more than one mechanism or pathway. Therefore, it is customary to use a combination of stabilizers because of a so-called synergistic effect. This term is used to explain the overall stabilization noted with a combination of stabilizers. The effect is greater than that predicted for a summation of the effects of each component.

Attempts to use stabilized rigid poly(vinyl chloride) prior to World War II were usually unsuccessful. American polymer chemists and engineers solved the problems associated with brittle PVC by plasticization and copolymerization. This will be discussed further in Secs. 10.2 and 10.3. In contrast, German chemists, through control of particle size, molecular weight distribution, stabilization, and control of processing temperature, were able to produce a useful rigid PVC.

This art, transported to the USA after cessation of hostilities, included the use of lubricants such as zinc stearate, pigments, fillers and heat stabilizers. In some instances about 5% of a plasticizer or softener was added to aid the processing, but this amount of plasticizer actually increases rather than decreases brittleness at ordinary temperatures. A polymer produced by the emulsion process is preferred for use as rigid PVC.

The additives are mixed with rigid PVC in a heavy-duty mixer and the temperature of the compounded plastic is kept at least 25°C below the decomposition temperature at all times. This plastic may be screw-injection molded, extruded by multiple-screw extrusion, calendered, and blow molded.

Approximately one billion pounds of rigid PVC are used annually in the USA as pipe, conduit, corrugated building panels, siding, gutters, downspouts, chemical processing equipment, phonograph records, and bottles. Complex structures may be fabricated by heat-forming rigid PVC sheet and heat-welding sections together by use of a heat gun and extruded PVC welding rod. Sheet or pipe may also be joined by the so-called *solvent welding process* (SWP) in which a viscous solution of PVC is applied before pressing the two surfaces together.

The properties of a typical molded *Type I rigid PVC* specimen are as follows: Specific gravity 1.4, tensile strength 7,000 psi, flexural strength 13,000 psi, compressive strength 10,000 psi, impact resistance 0.5 ft lbs/in. of notch, Rockwell hardness M115, coefficient of expansion 7×10^{-5} in./in./°C, heat-deflection temperature at 264 psi, 150°F, good electrical properties, good resistance to oils and ethanol, and excellent resistance to acids.

Relatively poor resistance to elevated temperature, such as that of household hot water, has been improved by chlorinating rigid PVC. This plastic has also been called poly(vinyl dichloride) (PVDC). A typical molded part of this plastic would have the following properties: Specific gravity 1.55, tensile strength 8,500 psi, flexural strength 16,000 psi, compressive strength 15,000 psi, impact strength 1.0 ft lbs/in. of notch, Rockwell hardness R120, coefficient of expansion 7.0×10^{-5} in./in./°C, heat-deflection temperature at 264 psi 220°F, good electrical properties, fair resistance to solvents, and good resistance to acids.

10.2 PLASTICIZED POLY(VINYL CHLORIDE)

As mentioned earlier, one technique used to improve the impact resistance of PVC is plasticization. This technique was used by Parkes in the United Kingdom in 1862 and by Hyatt in the USA in 1868. They added camphor to cellulose nitrate to produce an impact resistant plastic which was called Parkesine in Great Britain and Celluloid in the USA.

Semon added tricresyl phosphate [$(H_3CC_6H_4O)_3P{=}O$] to PVC and produced a flexible product called Koroseal in 1930. As pointed out in Sec. 3.3, these additives are called *external plasticizers*. Over a billion pounds of liquid plasticizers are used annually in the USA. Their major use is for the production of flexible PVC. The principal plasticizer is dioctyl phthalate (DOP). However, because of plasticizer migration and concern with the biological effects of phthalate esters, there is a trend toward the increased use of nonvolatile polymeric plasticizers.

The amount of plasticizer present in a flexible PVC article may range from 30–80%, depending on the efficiency of the plasticizer and the amount of flexibility desired. Additives, such as stabilizers, lubricants, pigments, and fillers, are incorporated in plasticized PVC. This compound plastic may be extruded, injection molded, calendered, and used as a solution or latex for protective coatings and films.

Plasticized poly(vinyl chloride) is used for making tubing, packaging, bottles, shoe soles and uppers, metal coatings, coatings for wire and cable, upholstery, wall covering, shower curtains, clothing, and electrical insulation tape.

A typical unfilled molded plasticized PVC article will have the following properties: Specific gravity 1.25, tensile strength 2,500 psi, Shore A hardness 75, coefficient of expansion 1.5×10^{-4} in./in./°C, good resistance to flame, good electrical properties, good resistance to oils and ethanol, and good resistance to acids.

Semirigid PVC, sometimes called *Type II rigid PVC*, is produced by using relatively-small amounts of liquid plasticizer or by blending with poly(methyl acrylate), butadiene-acrylonitrile copolymer elastomers, chlorinated polyethylene, or ABS copolymers. The products are more ductile than Type I rigid PVC and the ductility or flexibility is a function of the concentration of the polymeric additive. A typical molded article of ABS-modified PVC will have the following properties: Specific gravity 1.35, tensile strength 6,000 psi, flexural strength 7,500 psi, impact strength 5 ft lbs/in. of notch, Rockwell hardness R106, heat-deflection temperature of an annealed specimen at 264 psi, 160°F, good electrical properties, good resistance to mineral oils and ethanol, and good resistance to acids.

PVC plastisols, also called *pastes*, are actually dispersions of small PVC spheres in liquid plasticizers. These products were introduced commercially in 1937. They are usually prepared by dispersing PVC particles obtained by emulsion polymerization in a liquid plasticizer in a dough mixer. Dispersions of stabilizers, pigments, and fillers are also added in the mixer. When organic solvents are added, the product is called an *organosol*. When thickening agents such as pyrogenic silica or bentonite clay are added, the product is called a *plastigel*. Plastisols with moderate amounts of plasticizer yield semirigid products that are called *rigisols*.

Any of these dispersions may be used for coating paper, wire, cloth, or leather, for casting of articles, or for the production of foams. In all cases, the dispersions are heated at about 160°C. The plasticizer and the resin fuse or gel at this temperature and a homogeneous plasticized PVC is produced. Racks or tool handles are coated by dipping in a plastisol, draining and heating at 160°C. Gloves are produced by stripping the fused coating from the form. Balls and doll heads are produced by rotational casting at 160°C followed by the removing of excess plastisol.

Both organosols and solutions of PVC and plasticizers are used as coatings. Lattices produced by emulsion polymerization of vinyl chloride may also be used for coatings. Since the plasticizer does not usually penetrate into the polymer particles, it is necessary to heat emulsion coatings at 160°C in order to obtain homogeneous, plasticized coatings.

10.3 COPOLYMERS OF VINYL CHLORIDE

In 1928, chemists in the USA and in Germany produced copolymers of vinyl chloride and vinyl acetate $\left(\begin{array}{c} \text{H} \\ | \\ \text{H}_2\text{C}\!\!=\!\!\text{C}\!\!-\!\!\text{OOCCH}_3 \end{array}\right)$ which were more flexible and more readily processed than the homopolymers. The American product developed by Reid was called Vinylite and letters were used to designate the composition. For example, Vinylite VYNW and VYHH contained 3% and 13% of vinyl acetate, respectively.

The solution polymerization technique is preferred for the production of these copolymers. The term internal plasticization is sometimes used to express the flexibilizing effect of the vinyl acetate comonomer. These copolymers are more soluble than PVC in ester and ketone solvents. They are also less resistant to corrosives.

Low molecular-weight copolymers are used as coatings. Higher molecular-weight copolymers are used for extrusion and injection molding. The compounding recipes for the copolymer are similar to those used for PVC, but less plasticizer is required for the copolymers.

These copolymers may be processed at lower temperatures than PVC. They are used for upholstery, luggage, gaskets, coatings, insulation, flooring, and for many of the applications discussed previously for PVC. The properties of molded articles of this copolymer are similar to those listed for PVC but will be related to the composition of the copolymer and to the plasticizer content.

A copolymer of vinyl chloride and vinyl isobutyl ether

$$\left(\begin{array}{c} \quad\quad\quad \text{H} \quad \text{H}_2 \quad \text{H} \\ \quad\quad\quad | \quad\quad | \quad\quad | \\ \text{H}_2\text{C}\!\!=\!\!\text{COC}\!\!-\!\!\text{C}(\text{CH}_3)_2 \end{array}\right)$$

is produced commercially in Germany under the trade name of Vinoflex. The compounding recipes, processing, and end-uses of this copolymer are similar to those used for the vinyl acetate copolymer.

A copolymer of vinyl chloride and acrylonitrile $\left(\begin{array}{c} \text{H} \\ | \\ \text{H}_2\text{C}\!\!=\!\!\text{CCN} \end{array}\right)$ was introduced commercially under the trade name of Vinyon in 1947. A fiber called Dynel was produced in 1951 by extruding i.e., spinning an acetone solution of this copolymer.

Copolymers of vinyl chloride and diethyl maleate are produced in the USA under the trade name of Pliovic A. Copolymers of vinyl chloride and acrylic esters have been produced in Germany under the trade name of Mipolam. The vinyl acetate copolymers have been modified by saponifying some of the acetate groups in the copolymer to produce hydroxyl groups.

In another modification, maleic anhydride is included in the reactants. The resulting terpolymer has better adhesion than the copolymer.

A transparent copolymer that may be blow molded to form bottles is produced by the copolymerization of vinyl chloride and propylene. A typical article molded from this copolymer will have the following properties: Specific gravity 1.35, tensile strength 6,500 psi, flexural strength 13,000 psi, compressive strength 10,000 psi, impact strength 3 ft lbs/in. of notch, Rockwell hardness R112, heat deflection temperature at 264 psi, 160°F, good electrical properties, good resistance to mineral oil and ethanol, and excellent resistance to acids.

Copolymers of vinyl chloride and vinylidene chloride, in which the latter is the minor component, have been commercially available since 1938. The physical properties of these higher density copolymers are similar to copolymers of vinyl acetate or vinyl isobutyl ether. However, the corrosion resistance of the copolymer of vinyl chloride (90)-vinylidene chloride (10) is superior to a copolymer with 10% vinyl acetate.

Copolymers of vinyl chloride and vinyl bromide also have been produced in limited quantities. These copolymers have a higher specific gravity than PVC and resemble the vinylidene chloride copolymers.

10.4 POLYMERS OF VINYLIDENE CHLORIDE

Vinylidene chloride ($H_2C= CCl_2$) polymerizes readily in the presence of free radicals to produce a crystalline, insoluble, intractable polymer that softens at about 200°C and decomposes at about 210°C. This polymer is not readily plasticized, but, as stated in Sec. 10.3, does copolymerize with vinyl chloride. Wiley produced copolymers in which the major component was vinylidene chloride in the late 1930s.

These copolymers were introduced commercially under the trade name of Saran in 1937. The most widely-used product is a copolymer of 85% vinylidene chloride and 15% vinyl chloride. Stabilized Saran (PVDC) may be injection molded and extruded. It is of interest to note that Saran plastic pipe and fittings were commercially available before World War II and prior to rigid poly(vinyl chloride) pipe. However, because of its high specific gravity, the former is more expensive than other types of general purpose pipe and is not economically competitive.

Because of its negligible water vapor transmission, saran film is widely-used as a moisture barrier and as a heat shrinkable packaging film. The biaxially stretched film is called Saran Wrap. Saran filaments have been used for automobile seat covers, filter cloths, cordage, brushes, screens, and webbing. Copolymers of vinylidene chloride and vinyl chloride are used in the form of solutions and aqueous dispersions to produce coatings.

Approximately 100 million pounds of this plastic are produced annually in the USA and Japan. A typical part molded from this copolymer will have the following properties: Specific gravity 1.7, refractive index 1.61, tensile strength 10,000 psi, elongation 10%, impact strength 0.5 ft lbs/in. of notch, Rockwell hardness M85, fair electrical properties and excellent resistance to acids.

A copolymer of vinylidene chloride and acrylonitrile was introduced commercially by Dow in 1943. These copolymers are marketed under the trade name Saran F, and numbers are used to designate the ratio of the comonomers in the copolymer. The most widely-used product contains 15% acrylonitrile. A 50:50 vinylidene chloride-acrylonitrile copolymer is used as a fiber.

Saran F copolymers are used as solutions and aqueous dispersions for the formation of barrier coatings. The moisture vapor transmission and the solubility in ketones of these copolymer films increase as the acrylonitrile content increases. Saran F is often used as a coating for polyethylene films. A modification of this copolymer is also used as an elastomeric tank lining.

10.5 FLUOROPLASTICS

Prior to 1938, tetrasubstituted ethylenes, such as tetraflnoroethylene ($F_2C\!\!=\!\!CF_2$), were considered to be nonpolymerizable. However, in 1938 Plunkett noted the absence of pressure in a tank containing gaseous tetrafluoroethylene. He recovered the solid polymer by cutting open the tank and assigned the patent on this polymer to DuPont. It is of interest to note that all attempts to synthesize polytetrachloroethylene have been unsuccessful. The structure of the fluoropolymer, which is called Teflon and Fluon, is shown in Fig. 10-3.

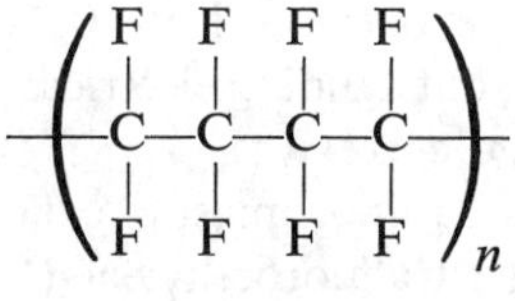

$$\left(\begin{array}{cccc} F & F & F & F \\ | & | & | & | \\ -C-C-C-C- \\ | & | & | & | \\ F & F & F & F \end{array} \right)_n$$

Fig. 10-3 A segment of polytetrafluoroethylene.

Teflon (TFE) was produced by free-radical polymerization on a small scale in 1941 and a commercial plant has been in operation since 1950. This polymer is available as a dispersion for coatings, and impregnants, and as granules for molding and extrusion. Processing aids must be used in processing, and the final product must be sintered at about 327°C in a manner similar to that used in powdered metallurgy. Teflon tape is produced by shaving molded or extruded articles.

As shown in Fig. 10-3, polytetrafluoroethylene is a linear polymer. Because of the stable carbon-fluorine bond and the closeness of the carbon atoms to each other, this polymer has outstanding resistance to heat. Because of the weak intermolecular forces characteristic of organic fluorides, polytetrafluoroethylene has a very low coefficient of friction. And because of the symmetry of its structure, this polymer is crystalline and does not melt below 327°C.

Polytetrafluoroethylene is soluble in highly fluorinated kerosene at temperatures above 300°C, but it is insoluble in all solvents at ordinary temperatures. It is attacked by molten sodium and by fluorine at room temperature but is inert to all other corrosives. It is customary to react the surface of Teflon films with a solution of sodium metal in liquid ammonia in order to alter the surface so that it may be adhered to other surfaces by epoxy resin adhesives. Films of this polymer retain their flexibility at temperatures as low as $-100°C$ and also retain their mechanical properties at temperatures up to 300°C. Thus, this plastic was used successfully during World War II for protecting equipment used for the separation of hexafluorides of isotopes of uranium.

Polytetrafluoroethylene (PTFE), in the form of film, filaments, pipe, tape, and molded articles, or pieces machined from billets, is used as wire insulation, gaskets, seals, stop cocks, packings, bearings, and corrosion-resistant coatings. This plastic may be filled with fibrous glass, graphite, asbestos, or metal powders to produce harder composites that may be used as pistons or other moving parts. Finely-divided Teflon is added to other plastics to provide good lubricity. Microporous Teflon is used as a filter in chemical processing plants.

The properties of a typical part molded from polytetrafluoroethylene are as follows: Specific gravity 2.2, tensile strength 3,500 psi, compressive strength 1,700 psi, impact strength 3.0 ft lbs/in. of notch, Shore D hardness 60, coefficient of thermal expansion 1×10^{-4} in./in./°C, heat deflection temperature at 66 psi 250°F, outstanding electrical properties, and outstanding resistance to all solvents and corrosives.

A more readily processed fluoroplastic, called *Teflon FEP*, is produced by the copolymerization of tetrafluoroethylene ($F_2C{=}CF_2$) and hexafluoropropylene $\left(F_3C-\underset{\underset{F}{|}}{C}{=}CF_2 \right)$. As illustrated in Fig. 10-4, this copolymer resembles the highly fluorinated PTFE shown in Fig. 10-3, but FEP has a trifluoromethyl pendant group (F_3C-) which permits greater chain mobility. Thus, FEP has a lower melting point and is more readily processed than PTFE. The properties and applications of this thermoplastic are similar to those cited previously for PTFE.

A more readily processed fluoroplastic has also been produced by copolymerizing ethylene and tetrafluoroethylene. This injection-moldable

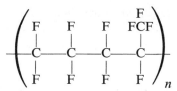

Fig. 10-4 A segment of fluorinated ethylene-propylene polymer (FEP).

melt-extrudable plastic was introduced commercially by DuPont in pilot-plant quantities in 1970 and in large-scale quantities in 1972. This copolymer, called *ETFE fluoroplastic*, is sold under the trade name Tefzel. Its dimensional stability can be enhanced by reinforcing with fibrous glass.

The properties of a typical part molded from ETFE are as follows: Specific gravity 1.7, tensile strength 6,500 psi, Rockwell hardness R50, coefficient of thermal expansion 4.2×10^{-5} in./in.°C, heat-deflection temperature at 66 psi 220°F, dielectric constant 2.6, other electrical properties outstanding, and excellent resistance to solvents and corrosives.

Trifluoromonochloroethylene ($F_2C{=}CFCl$) was polymerized in Germany in 1934. This polymer, called *CTFE*, was produced in limited quantity under the trade name Kel F prior to and during World War II. It was used for gaskets and valve seats for fluorine service in the diffusion process for separating the hexafluorides of the isotopes of uranium. It was introduced commercially by M. W. Kellogg Company in 1947 in the USA and is also sold in the USA under the trade names of Halon and Fluorethane. It had been marketed earlier in Germany under the name Hostaflon C2.

Because of the irregularity in structure caused by the introduction of a chlorine atom on every other carbon atom, CTFE is less crystalline than TFE. It is also more readily-molded and extruded than TFE and is used for similar applications.

An alternating copolymer of ethylene and trifluoromonochloroethylene (E-CTFE) is produced under the trade name of Halon. This readily processed plastic retains its useful properties from cryogenic temperatures to 325°F. Molded or extruded shapes of E-CTFE may be crosslinked by radiation. The properties of this copolymer are slightly inferior to those of CTFE and TFE but are superior to those of most other plastics.

A polymer of vinylidene fluoride ($H_2C{=}CF_2$) was introduced by Pennwalt Company in 1961 under the trade name of Kynar. This plastic, which is also called by the Symbol PVF_2, is readily extruded and injection molded. It is used as shrink tubing on electrical equipment, for chemical process piping, as a coil coating, and in porous form for fine line marking pens.

The properties of a typical article molded from PVF_2 are as follows: Specific gravity 1.76, tensile strength 6,500 psi, compressive strength 8,600 psi, impact strength 3.8 ft lbs/in. of notch, Shore D hardness 80, coefficient of thermal expansion 8.5×10^{-5} in./in./°C, heat-deflection temperature at 66 psi, 300°F, excellent electrical properties, and excellent resistance to most solvents and corrosives.

Monomeric vinyl fluoride $\left(H_2C\!=\!\overset{H}{\underset{}{C}F}\right)$ was prepared by Swarts in 1901 but prior to 1945, chemists considered this monomer to be unpolymerizable. Newkirk polymerized vinyl fluoride in the presence of benzoyl peroxide and water at 80°C and at a pressure of 1,000 atmospheres in 1946. This polymer was produced commercially in the early 1960s under the trade name of Tedlar.

Because of the small size of the fluorine atom, the PVF molecules are packed close together and the plastic has a high degree of crystallinity. PVF has better resistance to heat and weather than PVC. Its principal application is as a film used for electrical equipment, for agricultural glazing, and as a weather-resistant cladding for buildings.

A fluorine-containing elastomer obtained by the polymerization of 2-fluorobutadiene $\left(H_2C\!=\!\overset{F}{\underset{}{C}}\!-\!\overset{H}{\underset{}{C}}\!=\!CH_2\right)$ was introduced commercially under the trade name of fluoroprene in 1948. However, the high cost of this product hampered its acceptance.

Polymers of fluorinated acrylates, such as 1,1-dihydroperfluorobutyl acrylate $\left(H_2C\!=\!\overset{H}{\underset{|}{C}}\!-\!\overset{O}{\underset{\|}{C}}\!-\!O\!-\!\overset{H_2}{\underset{|}{C}}\!-\!C_3F_7\right)$, were also introduced commercially under the trade name of poly FBA, but their production has been abandoned because of brittleness at $-20°C$. However, copolymers of hexafluoropropylene (30) and vinylidene fluoride (70), which are produced under the code name HFP-VDF and the trade names of Viton and Fluorel, are available commercially.

The elastomeric HFP-VDF has better heat-aging and oil-resistance properties than the polymers of fluorinated acrylates but it also has poor low temperature properties. This oil resistant elastomer is used as a coating for asbestos for fire seals on aeroplane wings, and for o-rings and valve seats used in organic solvent service.

10.6 CHLORINATED POLYETHER

Corrosion resistant polymers were produced in the early 1950s by Hercules Powder Company and Imperial Chemical Industries by the cationic polymerization of 3,3-bis(chloromethyl)-1-oxacyclobutane. A segment of this polymer, marketed under the trade name Penton, is shown in Fig. 10-5. These polymer are usually called chlorinated polyethers.

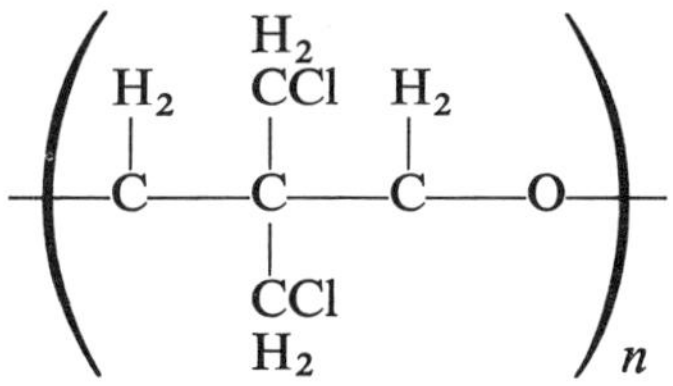

$$\left(\begin{array}{ccc} & H_2 & \\ H_2 & CCl & H_2 \\ | & | & | \\ -C- & C- & C-O- \\ & | & \\ & CCl & \\ & H_2 & \end{array} \right)_n$$

Fig. 10-5 A segment of a chlorinated polyether chain.

The crystalline linear polymer may be extruded or injection molded to produce articles that have excellent resistance to elevated temperature, solvents, and corrosives. Penton is used as a protective coating and lining as pipe and chemical processing equipment. The properties of a typical part molded from chlorinated polyether are as follows: Specific gravity 1.4, tensile strength 6,000 psi, flexural strength 5,000 psi, impact strength 0.4 ft lbs/in. of notch, Rockwell hardness R100, coefficient of thermal expansion 3.0×10^{-5} in./in./°C, heat deflection temperature at 66 psi 285°F, excellent electrical properties, and excellent resistance to solvents and corrosives.

BEHAVIORAL OBJECTIVES

Prior to the development of problems related to the toxicity of its monomer, PVC was one of the most widely used synthetic plastics. While some other materials can be used as substitutes for PVC, none can duplicate the characteristics of this important polymer.

After reading this chapter, you should understand the following concepts:

1. Most synthetic plastics are polymers of vinyl monomers, but the term vinyl plastic usually connotes PVC or copolymers of VCM.

2. Since PVC is decomposed by heat and ultraviolet light, it is customary to add appropriate stabilizers when compounding PVC molding or extrusion resins.

3. The heat resistance of PVC is improved by chlorination. The composition of the product is similar to that of poly(vinylidene chloride).

4. Prior to World War II, it was not possible to process PVC because the processing temperature was above the degradation temperature. Hence the large scale use of PVC is based on its plasticization by nonvolatile liquids such as dioctyl phthalate.

5. Copolymers of vinyl chloride and monomers such as vinyl acetate are more flexible than PVC. They may be processed at lower temperatures than the homopolymer and require less plasticizer for attainment of similar flexibility.

6. A crystalline polymer with a high density is obtained by the polymerization of vinylidene chloride.

7. A water and grease-repellent solvent and heat resistant polymer is obtained by the polymerization of tetrafluoroethylene. Polymers that are more readily processed are obtained by replacing one or more of the fluorine atoms in the monomer by other atoms or groups.

8. A heat and solvent resistant crystalline polymer is obtained by the polymerization of a chlorinated cyclic monomer called 3,3-bis-(chloromethyl)-1-oxacyclobutane. The polymer is called a chlorinated polyether.

GLOSSARY

DOP: Dioctyl phthalate.
Homopolymer: a polymer of identical monomer molecules.
Organosol: a plastisol containing solvent as well as liquid plasticizer.
Plasticizer: an additive which improves the impact resistance of a polymer.
Plastisol: a suspension of PVC in a liquid plasticizer.
PTFE: polytetrafluoroethylene.
PVC: poly(vinyl chloride).
PVCAC: poly(vinyl chloride-co-vinyl acetate).
PVDC: chlorinated PVC (poly(vinyl dichloride)).
Saran: a polymer produced by the polymerization of vinylidene chloride.
Semon, Waldo: the discoverer of plasticized PVC.
Shore: a measure of hardness.
Sintering: a process for fusing powdered polymer particles by heat and pressure.
Saponification: a process in which ester groups are hydrolyzed to produce a salt of the carboxylic acid and an alcohol, i.e., hydrolysis by alkaline solutions.
SWP: solvent welded pipe.
TCP: tricresyl phosphate.
Type II PVC: a semi-rigid PVC with improved resistance to impact.
VCM: vinyl chloride monomer.
Vinyl plastic: a term that usually connotes PVC or a copolymer of vinyl chloride.
Vinylite: trade name for a copolymer of vinyl chloride and vinyl acetate.

QUESTIONS

1. What is a vinyl plastic?
2. PVC was one of the first synthetic polymers, but it was not available commercially until the 1930's. Why?
3. What is the advantage of chlorinated PVC (PVDC)?
4. What is the function of a plasticizer?
5. What is the advantage of a copolymer of VCM and vinyl acetate?
6. Why is a PVC film more economical than a poly(vinylidene chloride) film?
7. Why does PTFE repel grease and water?
8. Why is the use of chlorinated polyether limited?

ANSWERS

1. Most thermoplastics are vinyl plastics, but this term usually connotes PVC or copolymers of VCM.
2. In the absence of plasticizers and stabilizers, PVC decomposes at processing temperatures.
3. PVDC is more heat resistant than PVC.
4. It reduces the glass transition temperature so that PVC becomes flexible and may be processed at lower temperatures than rigid PVC.
5. It has a lower glass transition temperature than PVC, is more flexible, and can be processed at lower temperatures.
6. Because of its lower density, an equal weight of PVC will produce a larger volume of film than PVDC.
7. It has a solubility parameter value which is lower than those of grease or water.
8. High cost of raw material and processing.

chapter 11

Saturated Polyesters

The term *polyester plastic* is often used nontechnically to describe fibrous glass reinforced composites made by the copolymerization of an unsaturated polyester and styrene. These and other thermosetting polyesters were discussed in Secs. 7.4, 7.5, and 7.7. There are also many thermoplastic polyesters such as cellulose nitrate, poly(ethylene terephthalate PETP) and polymers of difunctional or nonunsaturated esters. Cellulosics will be discussed in Chap. 14. Polyester fibers such as poly(ethylene terephthalate) (Dacron) will be mentioned in this chapter, but the emphasis will be on films (Mylar) and commercially important linear, thermoplastic polyesters.

11.1 POLYMETHACRYLATES

Acrylic acid $\left(H_2C{=}\overset{H}{C}{-}COOH\right)$ was synthesized in 1843, and the ethyl ester of methacrylic acid, i.e., ethyl methacrylate ($H_2C{=}C(CH_3)COOC_2H_5$), was prepared by Frankland and Duppa in 1865. Kahlbaum described poly(methyl methacrylate) (PMMA) in 1880. That the monomeric ester changed to a solid on standing was noted also by Fittig and Paul in 1900.

It is of interest to note that while his Belgium-born American contemporary Leo Baekeland was synthesizing phenolic resins, a German chemist named Rohm was synthesizing esters of acrylic and methacrylic acids. The Bakelite Company, established to produce phenolic resins, was acquired by Union Carbide Corporation in the early 1940s. The Rohm and Haas Company produced cast sheets of poly(methyl methacrylate) in the early 1930s and continues to produce and market acrylic resins today under the original names of Acryloid and Plexigum and under the trade name of Plexiglas.

Methacrylic acid is produced by the dehydration of α-hydroxyisobutyric acid. The latter is obtained by the nitric acid-nitrogen dioxide oxidation of isobutylene as shown in Fig. 11-1.

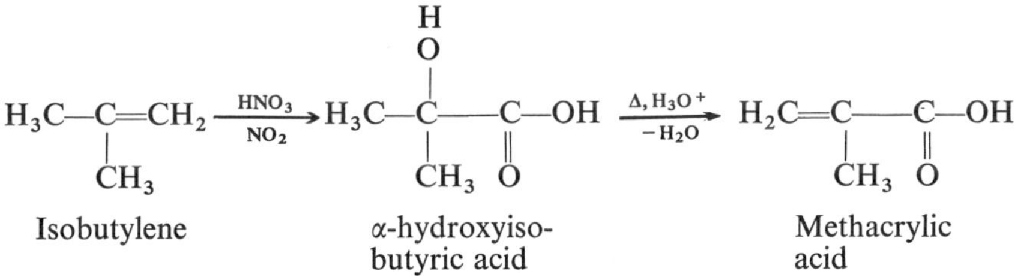

$$H_3C-\underset{CH_3}{\underset{|}{C}}=CH_2 \xrightarrow[NO_2]{HNO_3} H_3C-\underset{CH_3}{\underset{|}{\overset{\overset{\displaystyle H}{\overset{\displaystyle |}{O}}}{C}}}-\underset{O}{\overset{\|}{C}}-OH \xrightarrow[-H_2O]{\Delta, H_3O^+} H_2C=\underset{CH_3}{\underset{|}{C}}-\underset{O}{\overset{\|}{C}}-OH$$

Isobutylene α-hydroxyiso- Methacrylic
  butyric acid acid

Fig. 11-1 Synthesis of methacrylic acid.

The classical method for the synthesis of methyl methacrylate was by the methanolysis of acetone cyanohydrin. The latter was obtained by the hydrocyanation of acetone as shown in Fig. 11-2.

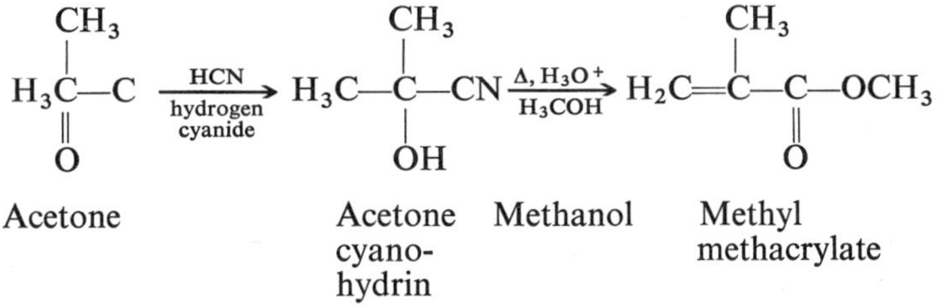

$$H_3C-\underset{O}{\overset{\|}{C}}-CH_3 \xrightarrow[\text{hydrogen cyanide}]{HCN} H_3C-\underset{OH}{\underset{|}{\overset{\overset{\displaystyle CH_3}{\overset{\displaystyle |}{}}}{C}}}-CN \xrightarrow[H_3COH]{\Delta, H_3O^+} H_2C=\underset{O}{\underset{\|}{\overset{\overset{\displaystyle CH_3}{\overset{\displaystyle |}{}}}{C}}}-C-OCH_3$$

Acetone Acetone Methanol Methyl
  cyano-   methacrylate
  hydrin

Fig. 11-2 Synthesis of methyl methacrylate.

Poly(methyl methacrylate) (Plexiglas, Lucite, Perspex) (PMMA) is obtained when the monomer is heated with organic peroxides in the absence of oxygen. The equation for this free radical (R·) polymerization is shown in Fig. 11-3.

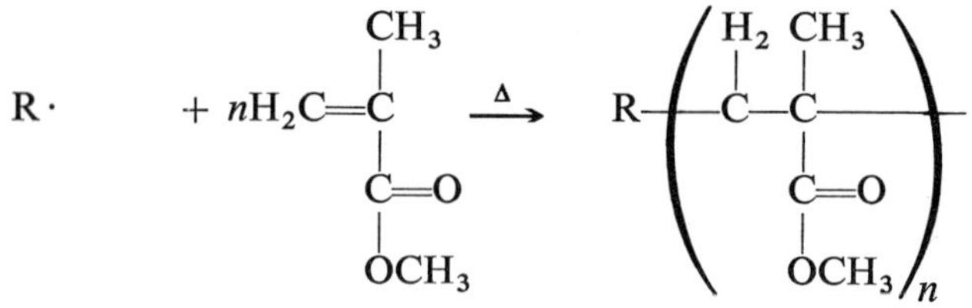

$$\text{R·} \quad + \quad n\text{H}_2\text{C}=\underset{\underset{\text{OCH}_3}{|}}{\underset{|}{\overset{\overset{\text{CH}_3}{|}}{\text{C}}}}\text{C}=\text{O} \quad \xrightarrow{\Delta} \quad \text{R}\!\!\left(\underset{\underset{\text{OCH}_3}{|}}{\underset{|}{\overset{\text{H}_2}{\text{C}}}}\!-\!\underset{\underset{\text{C}=\text{O}}{|}}{\overset{\overset{\text{CH}_3}{|}}{\text{C}}}\right)_{\!\!n}$$

Free radical from peroxide   Methyl methacrylate   Poly(methyl methacrylate) (PMMA)

Fig. 11-3 Polymerization of methyl methacrylate.

Methyl methacrylate may be bulk polymerized when castings, rods, or sheets are required. Large transparent sheets are made continuously by this technique. The casting technique is also used for embedding specimens and for potting. It is customary to partially prepolymerize the monomer to a syrup in a stirred vessel and to pour this syrup into the casting mold. Nitrogen or carbon dioxide is used to provide an anaerobic atmosphere.

The cast sheets may also be ground and used as molding powder. Considerable molding powder is produced by suspension polymerization. The emulsion polymerization technique is used when the product is to be applied as a coating or latex paint. The monomer is also polymerized in solution when a solution of polymer is required as the end product, such as automobile enamels.

Over 300 million pounds of acrylic plastics are produced annually in the USA. It is customary to add about 4% of dibutyl phthalate (DBP) as a plasticizer or to copolymerize the methyl methacrylate monomer with butyl acrylate $\left(\text{H}_2\text{C}=\overset{\overset{\text{H}}{|}}{\text{C}}-\text{COOC}_4\text{H}_9\right)$ in order to improve the impact resistance of poly(methyl methacrylate). Pigments, dyes, stabilizers, and fillers are incorporated in heavy-duty mixers or extruders when transparent products are not required.

Poly(methyl methacrylate) may be extruded or injection molded. The properties of cast or extruded sheets may be improved by biaxially stretching these sheets. The latter are readily thermoformed and may be joined by solvent welding techniques.

Clear poly(methyl methacrylate) sheets are used for residential, industrial, and aircraft glazing as well as for signs. Fibrous glass-reinforced corrugated sheets of this plastic are used for all types of construction. Cast and molded poly(methyl methacrylate) is used for dentures and wash basins. Molded poly(methyl methacrylate) is used for hair-brush backs, sign letters, and appliances.

The properties of a typical part molded from clear poly(methyl methacrylate) are as follows: Light transmission, more than 90%, specific gravity 1.2, index of refraction 1.49, tensile strength 9,000 psi, flexural strength 16,000 psi, compressive strength 15,000 psi, impact strength 0.4 ft lbs/in. of notch, Rockwell hardness M95, coefficient of thermal expansion, 7.0×10^{-5} in./in./°C, heat deflection temperature at 264 psi, 190°F, very good electrical properties, good resistance to mineral oil and ethanol, and good resistance to nonoxidizing acids.

As stated in Chap. 9, the resistance of this polymer to heat has been improved by copolymerizing the methyl methacrylate monomer with styrene or α-methylstyrene. These copolymers are marketed under the trade names of Implex, X-T Polymer, and Bavick. Elastomers are also blended with PMMA to improve the resistance to impact, and PMMA sheets are coated to protect them against scratches.

The softening point of poly(ethyl methacrylate) is less than that of PMMA, and this trend increases up to poly(dodecyl methacrylate). Polymers of esters of alcohols having higher molecular weights than dodecyl or lauryl alcohol have higher softening points because of an attraction between these longer ester groups. This increase in softening points is due to long side chain or pendant crystallization.

Polymers of butyl, octyl, and nonyl methacrylate are used in textile and leather finishes and in polishes. A solution of the copolymer of butyl and isobutyl methacrylate is used in aerosol bombs to produce artificial snow. Almost 50 million pounds of poly(dodecyl methacrylate) are used annually as a pour point depressant in lubricating oils. Soft contact lenses and dentures are produced by casting a crosslinked polymeric produced by the copolymerization of hydroxyethyl methacrylate ($H_2C=C(CH_3)-COO(CH_2)_2OH$)

and ethylene dimethacrylate $\left(\begin{array}{c} CH_3 \\ | \\ H_2C=C-COO(CH_2)_2OOC-C=CH_2 \\ | \\ CH_3 \end{array} \right)$.

11.2 POLYACRYLATES

As stated in Sec. 11.1, acrylic acid was synthesized 20 years before methacrylic acid. Actually the first commercial polymers marketed by Rohm and Haas in 1927 were solutions of acrylic ester polymers. Solutions of these polymers or copolymers are marketed under the trade name of Acryloid and the aqueous emulsions are known as Rhoplex resin emulsions.

Acrylic acid is produced by the nickel carbonyl ($Ni(CO)_4$)-catalyzed hydrocarbonylation ($CO + H_2O$) of acetylene or by the catalytic oxidation of propylene, as shown in Fig. 11-4.

Methyl acrylate may be produced by the methanolysis of acrylonitrile or ethylene cyanohydrin, as illustrated in Fig. 11-5. Other esters may be produced by substituting other alcohols for the methanol in these reactions. Esters of both acrylic and methacrylic acid may also be produced by transesterification of the methyl esters. In this reaction, a higher molecular weight

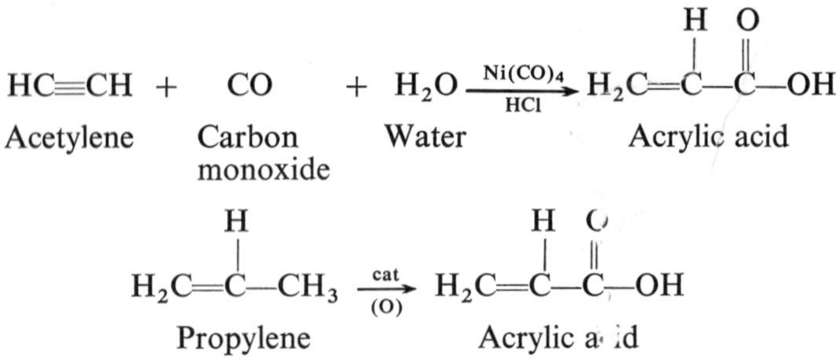

$$HC\equiv CH + CO + H_2O \xrightarrow[HCl]{Ni(CO)_4} H_2C=\underset{H}{\overset{}{C}}-\overset{O}{\overset{\|}{C}}-OH$$

Acetylene   Carbon monoxide   Water   Acrylic acid

$$H_2C=\underset{H}{\overset{}{C}}-CH_3 \xrightarrow[(O)]{cat} H_2C=\underset{H}{\overset{}{C}}-\overset{O}{\overset{\|}{C}}-OH$$

Propylene   Acrylic acid

Fig. 11-4 Synthesis of acrylic acid.

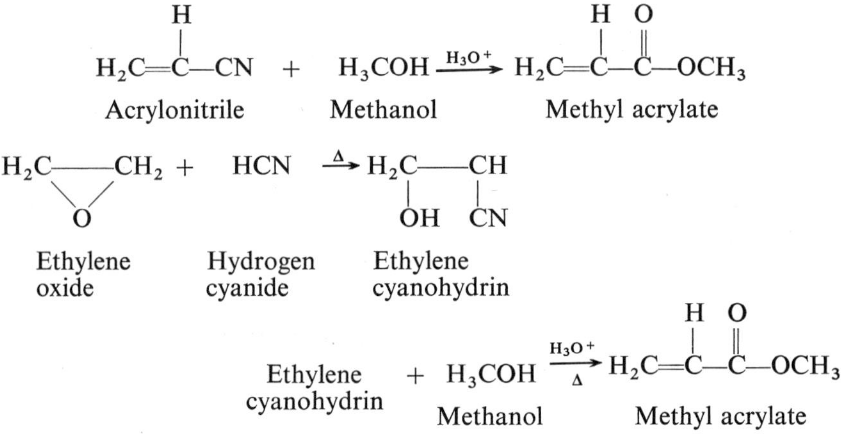

$$H_2C=\underset{H}{\overset{}{C}}-CN + H_3COH \xrightarrow{H_3O^+} H_2C=\underset{H}{\overset{}{C}}-\overset{O}{\overset{\|}{C}}-OCH_3$$

Acrylonitrile   Methanol   Methyl acrylate

$$H_2C\!\!-\!\!\!\underset{\underset{O}{}}{\!\!-\!\!}\!\!CH_2 + HCN \xrightarrow{\Delta} H_2C-CH$$
$$\qquad\qquad\qquad\qquad\qquad\quad \underset{OH}{|} \;\; \underset{CN}{|}$$

Ethylene oxide   Hydrogen cyanide   Ethylene cyanohydrin

$$\text{Ethylene cyanohydrin} + H_3COH \xrightarrow[\Delta]{H_3O^+} H_2C=\underset{H}{\overset{}{C}}-\overset{O}{\overset{\|}{C}}-OCH_3$$

Methanol   Methyl acrylate

Fig. 11-5 Synthesis of methyl acrylate.

alcohol replaces the methanol that is hydrolyzed off in the presence of sodium hydroxide removed by distillation.

Polymers of acrylic acid and its esters are produced by the free radical polymerization of the monomers. Acrylic acid may be polymerized in water to produce a viscous solution of poly(acrylic acid). The sodium and ammonium salts are more soluble, and aqueous solutions of these salts are used as thickeners, textile sizes, adhesives, and viscosity control agents. Poly(methacrylic acid, has similar properties to that of poly(acrylic acid).

Poly(methyl acrylate) is a tough leathery product. As was the case for the higher alkyl esters of methacrylic acid discussed in Sec. 11.1, the softening point of polyacrylates decreases as the molecular weight of the ester group increases up to the dodecyl ester. The softening point then increases as the size of the ester group increases because of side-chain crystallization.

Emulsions and solutions of homopolymers of acrylic esters are used as adhesives and coatings. The impact resistance of PMMA is improved by copolymerizing with acrylic esters. A copolymer of ethyl acrylate (95) and α-chloroethyl vinyl ether (5) has been used as an oil resistant elastomer. As mentioned in Chap. 8, copolymers of methyl acrylate and ethylene are available commercially. In contrast to poly(methyl methacrylate), which is resistant to alkalies, poly(alkyl acrylates) are readily saponified by aqueous sodium hydroxide.

In spite of the tear producing properties of its monomer, poly(methyl-α-chloroacrylate) has been used as a more expensive, higher softening, transparent plastic sheet. Methyl α-cyanoacrylate polymerizes anionically in the presence of moisture at room temperature. Since the polymer is an excellent adhesive, this relatively expensive monomer is used to adhere a wide variety of materials.

11.3 POLYCARBONATES

Esters of monohydric alcohols (ROH) and carbonic acid $\left(\text{HO}-\overset{\overset{\text{O}}{\|}}{\text{C}}-\text{OH}\right)$ have been known for over a century, but commercial polymers of carbonic esters of difunctional or dihydric alcohols are relatively new. It is of interest to note that Einhorn condensed phosgene $\left(\text{Cl}-\overset{\overset{\text{O}}{\|}}{\text{C}}-\text{Cl}\right)$ with hydroquinone $\left(\text{HO}-\!\!\bigcirc\!\!-\text{OH}\right)$ in 1898 but, like most good organic chemists, he discarded the product because it was an insoluble glassy material and not the low molecular weight cyclic ester that he was seeking. Bischoft and Von Hedenstrom also prepared polycarbonates early in 1902 but did not obtain useful products.

As pointed out in Sec. 7.7, the polymer of diethylene glycol bis(allyl carbonate) was introduced under the trade name of CR-39 in 1941. Carothers and Natta prepared linear aliphatic polycarbonates in 1930 but did not pursue the synthesis of these or other polyesters further because the aliphatic polycarbonates were readily hydrolyzed by water and none of the aliphatic polyesters were potential fibers.

The results of relevant research on carbonic acid esters of bisphenol A were announced in 1953 by Bayer $\left(\text{HO}-\!\!\bigcirc\!\!-\overset{\overset{(CH_3)_2}{|}}{\text{C}}-\!\!\bigcirc\!\!-\text{OH}\right)$ in Germany and by General Electric in the USA. Moldable linear polycarbonates (PC) were marketed in both countries under the trade names of Merlon and Lexan in 1957. The name Panlite is used to describe the Japanese product. As shown in the equation in Fig. 11-6, this engineering plastic may

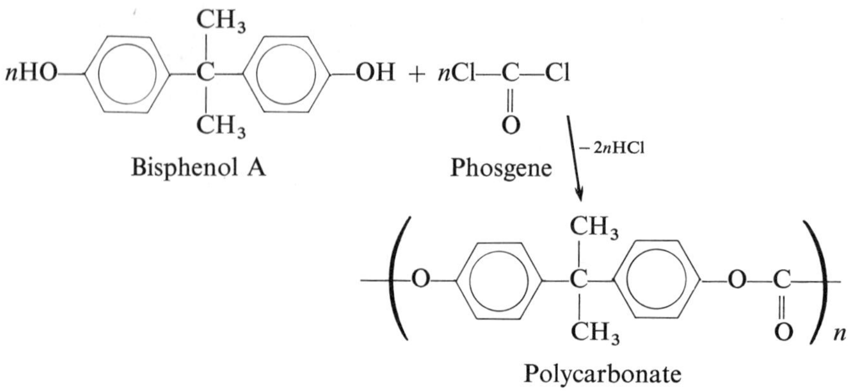

Fig. 11-6 Synthesis of a polycarbonate.

be produced by the condensation of phosgene ($COCl_2$) and bisphenol A. This thermoplastic is also prepared by the transesterification of diphenyl carbonate and bisphenol A in a manner similar to that outlined in Sec. 11.2 for the synthesis of esters of acrylic and methacrylic acid.

Dry polycarbonate resins may be compression or injection molded, cast, extruded, and blow molded. Unlike Carothers' aliphatic polycarbonates, the molded aromatic polycarbonates are resistant to hydrolysis by water but are hydrolyzed by strong aqueous alkaline solutions. Since the molding powder is degraded by hot water, it is essential that dry resin be used for thermal processing such as molding. Polycarbonate plastics tend to craze when exposed to solvents such as carbon tetrachloride. These plastics may be reinforced with fibrous glass and blended with ABS plastics.

Polycarbonates are used for interior components of aircraft, household appliances, lenses, switch gears, transformer housings, sports helmets, recreational vehicle bodies, propellers, body armor, and glazing. The properties of a typical molded unfilled transparent polycarbonate part are as follows: Light transmittance 89%, index of refraction 1.586, specific gravity 1.75, tensile strength 9,000 psi, flexural strength 15 ft lbs/in. of notch, Rockwell hardness R120, coefficient of expansion 6.6×10^{-5} in./in./°C, heat deflection temperature at 264 psi, 275°F, excellent electrical properties, good resistance to mineral oils and ethanol, fair resistance to weak aqueous acid solutions, and poor resistance to alkalies.

11.4 POLY(VINYL CARBOXYLATES)

The principal poly(vinyl carboxylate) is produced by the polymerization of vinyl acetate $H_2C{=}COCCH_3$. This monomer was prepared by Klatte in

1912 by the reaction of acetic acid and acetylene. Klatte also produced poly(vinyl acetate), which he proposed as a substitute for the more flammable cellulose nitrate (Celluloid). Poly(vinyl acetate) was produced commercially by the Shawinigan Resins Corporation in Canada in 1920. The trade names for this polymer are Elvacet and Gelva.

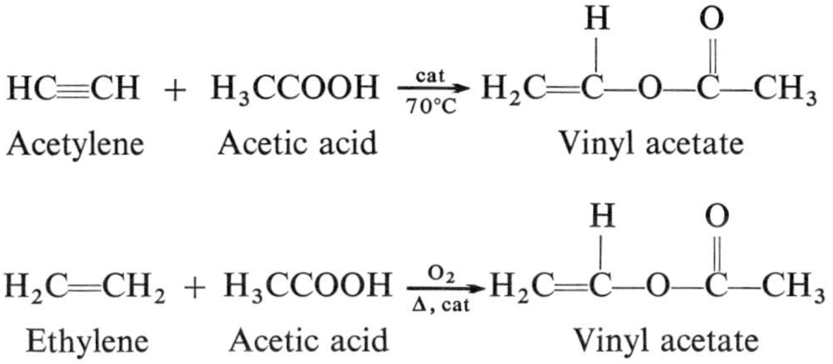

$$HC \equiv CH + H_3CCOOH \xrightarrow[70°C]{cat} H_2C=CH-O-C(=O)-CH_3$$
Acetylene + Acetic acid → Vinyl acetate

$$H_2C=CH_2 + H_3CCOOH \xrightarrow[\Delta, cat]{O_2} H_2C=CH-O-C(=O)-CH_3$$
Ethylene + Acetic acid → Vinyl acetate

Fig. 11-7 Synthesis of vinyl acetate.

Figure 11-7 shows that vinyl acetate is prepared both by the Shawinigan process discussed above and by the oxyacetylation of ethylene. Other vinyl esters are usually produced by transesterification of vinyl acetate with a higher boiling acid than acetic acid.

Vinyl acetate may be polymerized by free-radical polymerization to produce poly(vinyl acetate) (PVAC) by bulk, solution, suspension, or emulsion polymerization. The equation for this polymerization is shown in Fig. 11-8. The emulsion process is preferred when the product is to be used as a latex paint. However, the suspension process is preferred when the polymer is used for the preparation of derived polymers such as poly(vinyl alcohol) and poly(vinyl butyral).

Because of its low softening point, poly(vinyl acetate) is not used as an unfilled molding resin. The principal use of this polymer is in latex paints,

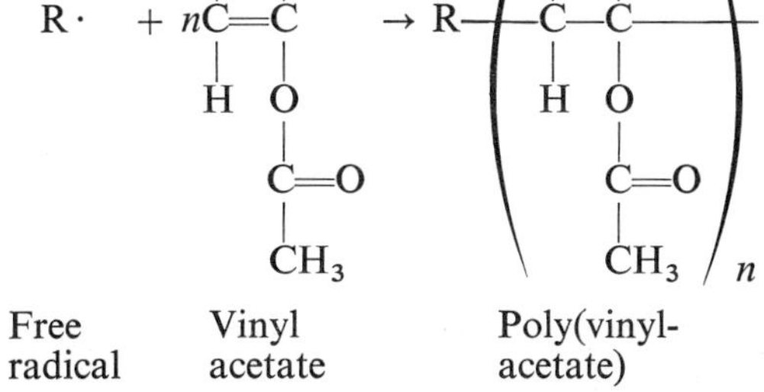

$$R\cdot + n\underset{\substack{| \\ H \quad O \\ | \\ C=O \\ | \\ CH_3}}{\overset{H \quad H}{C=C}} \rightarrow R-\left(\underset{\substack{| \\ H \quad O \\ | \\ C=O \\ | \\ CH_3}}{\overset{H \quad H}{C-C}}\right)_n$$

Free radical | Vinyl acetate | Poly(vinyl-acetate)

Fig. 11-8 Equation for the polymerization of vinyl acetate.

adhesives, and as a permanent starch. One of the principal uses of the monomer is as a comonomer for copolymerization with ethylene, vinyl chloride, or maleic anhydride. The latter copolymer was marketed as a soil conditioner under the trade name of Krilium. Polymers of other vinyl carboxylates are used for specialty applications, and the monomers are used as comonomers for copolymerization.

11.5 POLY(P-OXYBENZOATE)

A highly-crystalline, temperature-resistant polymer marketed under the trade name of Ekonol has been produced by the self-condensation of p-hydroxybenzoic acid (HO—⟨○⟩—COOH). This linear polymer chain is made up of repeating p-oxybenzoyl units (—O—⟨○⟩—C—).
$$\overset{\|}{O}$$

Because of its high melting point, which is greater than 900°C, this polymer cannot be processed by traditional extrusion or molding techniques.

Thus, this plastic is formed into useable objects by long time (1–5 min) compression molding at 350–425°C and a pressure of 5,000–20,000 psi. Sintering, like that observed in powdered-metal molding technology, occurs under these conditions. Because of the existence of thermal stresses in the sintered parts, it is customary to incorporate about 18% of a finely-divided alumina filler in the polymer before fabrication.

This polymer may also be flame sprayed and formed by high-energy forging. Because of its inherent self-lubricating qualities, this plastic is used for bearings, seals, rotors and vanes of process pumps. It has excellent electrical properties and is resistant to all solvents and corrosives.

11.6 POLYPHTHALATES

As discussed in Sec. 11.3, Carothers abandoned the investigation of polyesters because of the nonfibrous characteristics of high molecular weight aliphatic polyesters. Also, as discussed in Chap. 7, crosslinked polyester coatings have been produced from phthalic acid and glycerol since the early part of the 20th century. In spite of this background, which suggested that polyesters were nonfibrous, chemists at Imperial Chemical Industries were able to produce excellent fibers by the transesterification of dimethyl phthalate and ethylene glycol.

The poly(ethylene terephthalate) (PETP) produced was called Terylene. This was later produced in the USA under the trade names of Dacron and

Sec. 11.6 POLYPHTHALATES

Kodel. These polyester fibers are now being produced in the USA at an annual rate of more than one billion pounds. These PETP polymers were also extruded as a strong transparent film called Mylar.

Poly(ethylene terephthalate) film has a tensile strength of 25,000 psi, which is over 300% greater than that of films of cellophane, polyethylene, or cellulose acetate. Biaxially oriented PETP is widely used as magnetic, video, and industrial tape, microfilm, wire and cable wrapping, and packaging. It is customary to coat the packaging film with poly(vinylidene chloride-co-vinyl chloride) to reduce oxygen transmission. Metallized PETP film is used for decorative film, labels, and decals.

In spite of the excellent properties of PETP fiber and film, this polymer is not sufficiently ductile for use as a plastic. The difficulty in processing of PETP and its low ductility have been overcome by the production of phthalic acid esters with higher molecular weight dihydric alcohols such as butylene glycol $(HO(CH_2)_4OH)$. The difference in the structural formulas of PETP and the thermoplastic phthalic acid esters used for molding and extrusion are shown in Fig. 11-9.

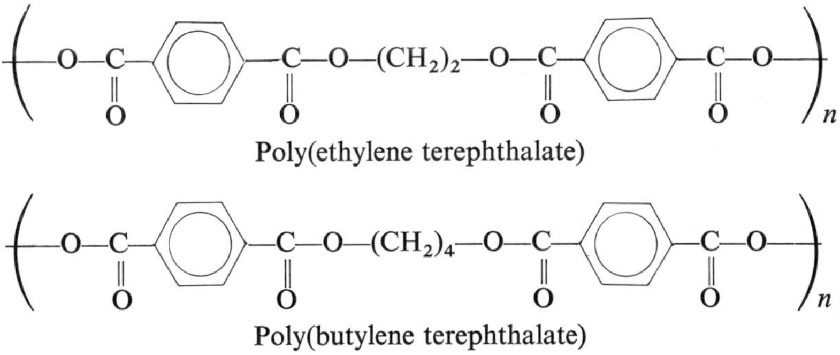

$$\left(-O-\underset{O}{\underset{\|}{C}}-\underset{}{\bigcirc}-\underset{O}{\underset{\|}{C}}-O-(CH_2)_2-O-\underset{O}{\underset{\|}{C}}-\underset{}{\bigcirc}-\underset{O}{\underset{\|}{C}}-O-\right)_n$$

Poly(ethylene terephthalate)

$$\left(-O-\underset{O}{\underset{\|}{C}}-\underset{}{\bigcirc}-\underset{O}{\underset{\|}{C}}-O-(CH_2)_4-O-\underset{O}{\underset{\|}{C}}-\underset{}{\bigcirc}-\underset{O}{\underset{\|}{C}}-O-\right)_n$$

Poly(butylene terephthalate)

Fig. 11-9 Structural formulas for aromatic polyesters.

The moldable thermoplastic aromatic polyesters are marketed under the trade names of Celanex, Tenite, and Valox. This highly-crystalline thermoplastic may be extruded or injection molded. It may be reinforced by fibrous glass and asbestos.

These engineering or high performance thermoplastics are used as gears, bearings, pump impellers, pulleys, pump housings, switch parts, and furniture parts. An amorphous transparent modification which may be sterilized without damage is available and is used for packaging. The properties of a typical part molded from a thermoplastic aromatic polyester are as follows: Tensile strength 8,000 psi, flexural strength 13,000 psi, compressive strength 13,000 psi, impact strength 1.2 ft lbs/in. of notch, Rockwell hardness R117, coefficient of thermal expansion 5.3 in./in./°C, heat deflection temperature at 66 psi 310°F, excellent electrical properties, and good resistance to solvents and acids.

BEHAVIORAL OBJECTIVES

While the term polyester plastic usually connotes fibrous glass-reinforced thermoset polyesters, several thermoplastics are also polyesters. Some of these are engineering plastics.

After reading this chapter, you should understand the following:

1. Poly(methyl methacrylate) is obtained by the free radical chain polymerization of methyl methacrylate. Objects such as sheet and rod may be obtained by polymerizing a prepolymer syrup of this monomer in a suitable mold.
2. Poly(dodecyl methacrylate) is oil soluble and is used as a pour point depressant in lubricating oils.
3. Acrylic ester polymers are more flexible and have lower softening points than the corresponding methacrylic esters.
4. Esters of alpha-cyanoacrylic acid polymerize readily in the presence of moisture to produce a product with excellent adhesive properties.
5. Polycarbonates, which are usually bisphenol A esters of carbonic acid, are transparent products with excellent impact resistance.
6. Poly(vinyl acetate), which is produced by the free radical polymerization of vinyl acetate, is used as the binder in latex paints and for the production of poly(vinyl alcohol).
7. A crystalline heat resistant polyester is produced by the self-condensation of p-hydroxybenzoic acid.
8. Heat resistant polyesters are also produced by the condensation of terephthalic acid and ethylene glycol. While this poly(ethylene terephthalate) (PETP) is preferred for polyester fibers, the condensation product of terephthalic acid and 1,4-butylene glycol is preferred for molded plastics.

GLOSSARY

Carbonic acid: a solution of carbon dioxide in water, i.e., soda water.
Condensation: the reaction of two molecules with functional groups, such as the preparation of an ester from an alcohol and an acid. When both functional groups are present in the same molecule, the process is called self-condensation.
Dihydric: having two hydroxyl groups.
Ester: product obtained by the condensation of an alcohol and a carboxylic acid.
Lachrymator: tear producer.
PETP: poly(ethylene terephthalate).
PMMA: poly(methyl methacrylate).
PVAC: poly(vinyl acetate).

QUESTIONS

1. How do you produce a transparent casting?
2. Which is more resistant to gasoline, (a) poly(methyl methacrylate) or (b) poly(dodecyl methacrylate)?
3. Which is more flexible (a) poly(methyl methacrylate) or (b) poly(ethyl acrylate)?
4. Why must cyanoacrylic esters be kept dry?
5. Which would you select for bullet proof glazing, (a) poly(methyl methacrylate), or (b) polycarbonate?
6. Why is poly(vinyl alcohol) produced by the saponification of poly(vinyl acetate) rather than by the polymerization of vinyl alcohol?
7. Can you suggest a use for polymers of p-hydroxybenzoic acid?
8. Which polyester is preferred for fibers?

ANSWERS

1. Heat a mixture of methyl methacrylate and benzoyl peroxide until a viscous syrup is produced. Then pour this syrup into a simple mold and heat until the liquid is converted to a solid.
2. (a).
3. (b).
4. To prevent uncontrolled polymerization.
5. (b). The polycarbonate has much greater resistance to impact.
6. Vinyl alcohol does not exist.
7. For molding articles or extruding pipe that are used in high temperature service.
8. PETP.

chapter 12

Polyamides, Polyimides and Polyurethanes

The polymers discussed in this chapter are all nitrogenous or nitrogen-containing polymers. While admittedly the first manmade plastic was ebonite, the plastic industry considers celluloid to be the first manmade plastic. This plasticized cellulose nitrate is, of course, nitrogenous as also is casein. Casein was also used as one of the pioneer plastics and adhesives and has also been used as a fiber.

Casein and all other proteins are polyamides. Molded casein articles, which had been surface hardened by immersion in formalin, were marketed under the trade name of Galalith in Germany in the early part of the 20th century. Cattle blood from slaughter houses is used as an industrial adhesive, and leather is produced by a reaction of the protein of cattle hides with crosslinking agents such as tannic acid. Vegetable oil polyamides are used as surface coatings.

Protein containing fibers, such as wool and silk, are polyamides that are used in their natural states without further chemical reactions. However, regenerated proteins, such as those from soybeans, peanuts, casein, and feathers, are usually reacted with formaldehyde after being spun. In spite of the importance of these biopolyamides, the emphasis in this chapter is on synthetic polyamides, polyimides and polyurethanes.

12.1 POLYACRYLAMIDE

Acrylamide $\left(\text{H}_2\text{C}=\overset{\text{H}}{\underset{|}{\text{C}}}-\overset{\text{O}}{\underset{\|}{\text{C}}}\text{NH}_2\right)$, which may be produced by the thermal dehydration of ammonium acrylate $\left(\text{H}_2\text{C}=\overset{\text{H}}{\underset{|}{\text{C}}}-\overset{\text{O}}{\underset{\|}{\text{C}}}-\text{O}^-,\text{NH}_4^+\right)$, is readily polymerized in the presence of free radicals to produce polyacrylamide

$$\left(\begin{array}{c} \text{H} \ \ \text{H} \\ | \ \ \ | \\ -\text{C}-\text{C}- \\ | \ \ \ | \\ \text{H} \ \ \text{C}=\text{O} \\ \ \ \ \ \ | \\ \ \ \ \ \text{NH}_2 \end{array}\right)_n$$

. This polymer may be applied as a coating and crosslinked with formaldehyde after application. Copolymers of acrylamide and styrene or methyl methacrylate have also been used as coatings and as sizings for paper and textiles.

The principal use of polyacrylamide is as a flocculent in sewage treatment and as a scale inhibitor in boiler water. The monomer may be polymerized in an aqueous solution using either potassium persulfate or ultraviolet light and oxygen as the initiator. The product is marketed as an aqueous solution under the trade name of Separan.

Small amounts of polyacrylamide, used as a flocculating agent in the presence or absence of alum, increase the efficiency of sewage treatment plants by as much as 200%. Presumably, the salts of polyvalent cations, such as calcium or aluminum, serve as high molecular-weight lakes that remove contaminants from water as the floc settles. It is also believed that this water-soluble polymer inhibits scale formation by interfering with the normal growth of crystals.

12.2 NYLON-66

As mentioned in Chap. 11, Carothers synthesized linear aliphatic polyesters in 1928 but abandoned this investigation in order to study polyamides. He prepared nylon-66 in 1935 by the thermal dehydration of the salt obtained by the reaction of adipic acid ($HOOC(CH_2)_4COOH$) and hexamethylenediamine ($H_2N(CH_2)_6NH_2$). Synthetic fibers were produced from this polyamide in 1939. The original product was called by the trade name of *nylon*, but nylon is now used as a generic name to describe all commercial synthetic high molecular-weight polyamides.

The numbers that follow the various types of nylon (PA) indicate the composition. Thus, the numbers 66 are used to show that the most widely-used type of nylon is produced by the reaction of a difunctional amine and a difunctional carboxylic acid in which each contains six carbon atoms. This polymer is marketed under the trade name of Zytel when used as a plastic. Over 1.5 billion pounds of nylon-66 are produced annually in the USA, but less than 125 million pounds of this polymer are used as plastics.

As shown in Fig. 12-1, nylon-66 salt, which has been purified by crystallization from ethanol, is heated to produce a high molecular-weight polymer. This is a general reaction that is also used for the production of nylon copolymers.

Since the molecular weight of a step-reaction polymer or condensation polymer is related to the purity of the reactants, nylons produced by heating pure salts have extremely-high molecular weights. After experiencing considerable difficulty in producing high molecular-weight polyesters, Carothers was so impressed with the high molecular-weight of nylon-66 that he

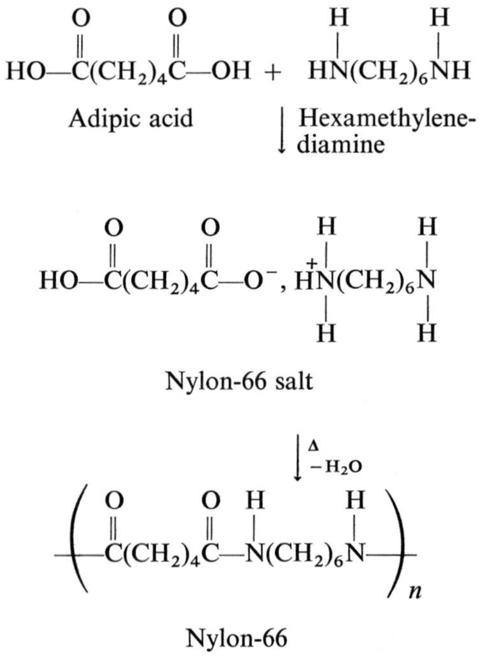

$$\text{HO—C(CH}_2)_4\text{C—OH} + \text{HN(CH}_2)_6\text{NH}$$
$$\underset{\text{Adipic acid}}{} \quad \underset{\text{Hexamethylene-diamine}}{}$$

$$\downarrow$$

$$\text{HO—C(CH}_2)_4\text{C—O}^-, \text{H}\overset{+}{\text{N}}(\text{CH}_2)_6\text{N}$$

Nylon-66 salt

$$\downarrow \overset{\Delta}{-\text{H}_2\text{O}}$$

$$\left(-\text{C(CH}_2)_4\text{C—N(CH}_2)_6\text{N}- \right)_n$$

Nylon-66

Fig. 12-1 Synthesis of nylon-66.

called it a *superpolyamide*. Actually, these extremely-high molecular-weight polymers are difficult to process and the molecular weight is reduced by adding a small amount of a monofunctional acid viz., acetic acid to the nylon salt.

The high melting point of nylon-66 may be lowered by copolymerization with another dicarboxylic acid and another diamine. When three different dicarboxylic acids and three different diamines are used, flexible terpolymers that are soluble in aqueous ethanol are obtained. Nylon-66 and other nylons may also be made more flexible by addition of plasticizers such as aromatic sulfonamides. It is customary to add lubricants or mold release agents when these nylons are molded or extruded.

The tendency for nylons to become brittle at temperatures above 175°F is reduced by the addition of stabilizers such as copper salts. Pigmented nylons are not affected adversely by exposure to sunlight, but ultraviolet stabilizers must be added to unpigmented molding powders. Since nylon moldings crystallize as they are cooled, some distortion of the molded part may occur. This effect may be prevented by incorporating nucleating agents that promote rapid crystal growth in the molding powders.

The physical properties of nylons are improved considerably by the addition of reinforcing agents such as fibrous glass or potassium titanate crystals. Molding quality is also improved by adding a small amount of glass spheres to the reinforced molding powder. Finely-divided polytetrafluoroethylene or molybdenum disulfide may be added to increase the lubricity of molded parts. It should be noted that all nylons are hygroscopic and thus exposure to moisture results in a lowering of some physical properties and improvement in others.

Because of its toughness and excellent resistance to abrasion, nylon-66 is used in many automotive and appliance parts, such as gears, bushings, cams, and bobbins. Nylon filament and bristles are used for fishing lines and brushes. Nylon-66 is also used as cast film, extruded sheet, tubing and rod, and as a jacketing and coating for wire and cable. Aerosol valves are usually molded from nylon.

The properties of a typical dry part molded from nucleated nylon-66 are as follows: Tensile strength 15,000 psi, compressive strength 15,000 psi, impact strength 0.9 ft lbs/in. of notch, coefficient of thermal expansion 8.0×10^{-5} in./in./°C, heat deflection temperature at 66 psi 220°F, Rockwell hardness R123, good electrical properties, good resistance to mineral oils, and poor resistance to acids. The improvement in properties obtained with 30% fibrous glass reinforcement is shown by the following data: Tensile strength 25,000 psi, flexural strength 38,000 psi, compressive strength 20,000 psi, impact strength 2.5 ft lbs/in. of notch, and coefficient of expansion 2.0×10^{-5} in./in./°C.

12.3 NYLON-6

Von Braun produced high molecular weight polyamides by the ionic polymerization of amino acids in 1907. Presumably the German chemists chose this route for the production of polyamides in order to circumvent the DuPont patents. Thus nylon-6 was produced from ε-aminocaproic acid in Germany during World War II. However, as shown in Fig. 12-2, this amino acid has been replaced as a reactant by its anhydride ε-caprolactam. Nylon-6 has been produced commercially in the USA since the 1950s by the polymerization of ε-caprolactam in the presence of water and a small amount of acetic acid.

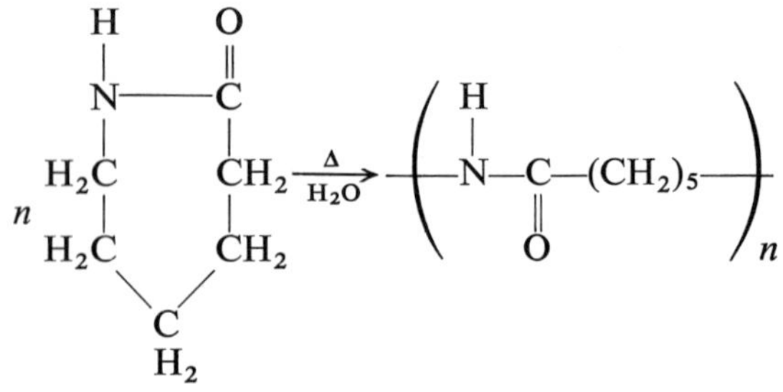

$$n \left[\text{ε-Caprolactam} \right] \xrightarrow[H_2O]{\Delta} \left(-N-C-(CH_2)_5- \right)_n$$

ε-Caprolactam   Nylon-6

Fig. 12-2 Equation for the synthesis of nylon-6.

Since this polymerization takes place readily without the formation of byproducts, ε-caprolactam may be polymerized *in situ* to produce large and complex parts which can not be readily produced by molding. For example, castings weighing as much as 1 ton have been produced by this *in situ* polymerization technique. However, it should be noted that the cast product contains about 10% of unreacted monomer which has an adverse effect on the physical properties. This residual monomer is removed by solvent extraction or vacuum distillation when the polymer is used as a molding powder.

The properties of a typical dry part molded from nucleated nylon-6 are as follows: Tensile strength 13,000 psi, flexural strength 17,500 psi, compressive, strength 16,000 psi, impact strength 0.9 ft lbs/in. of notch, Rockwell hardness R119, coefficient of thermal expansion 8.0×10^{-5} in./in./°C, heat deflection temperature 370°F, good electrical properties, good resistance to mineral oils, and poor resistance to acids. The properties of cast nylon-6 differ slightly from those of the molded plastic because of contamination with unreacted monomer. The properties of the molded fibrous glass-reinforced nylon-6 are improved in a manner comparable to that noted previously for nylon-66.

12.4 OTHER POLYAMIDES

The synthesis of polymers from nylon salts or amino acids is a versatile reaction and thus a large number of polymers have been prepared in the laboratory. It is important to note that the hygroscopicity of nylons increases and the melting point decreases as the number of methylene groups in the building units decrease and vice versa.

Thus nylon-610, produced by the condensation of hexamethylenediamine $(H_2N(CH_2)_6NH_2)$ and sebacic acid $(HOOC(CH_2)_8COOH)$, has a lower softening point and lower water absorption than nylon-66. Nylon-610 is preferred for brush bristles. Nylon-612 is made by the condensation of hexamethylenediamine and α,ω-dodecanoic acid $(HOOC(CH_2)_{10}COOH)$. Nylon-1313 has been produced by the U.S. Department of Agriculture but has not been produced commercially.

When nonlinear reactants are used, the irregularities in the structure of the resultant polymer retard crystallization. Thus, branched-chained nylons have lower melting points, and are more flexible and more transparent. Nylon-66 has a melting point of 265°C, but the branched nylons produced by the condensation of hexamethylenediamine and α-methyladipic acid as well as from the condensation of 3-methylhexamethylenediamine and adipic acid have melting points of 166°C and 180°C respectively. Nylons will react with formaldehyde in ethanol to yield N-alkoxymethyl substituted nylons that, like other branched nylons, have greater flexibility and lower melting points.

Copolymers have lower melting points than nylon homopolymers. Thus, the nylon-66/610 copolymer is more readily processed than either nylon-66 or nylon-610. The monofilament of this copolymer is preferred for fishing lines. Since the terpolymer nylon-66/610/6 is soluble in aqueous ethanol, it is used as a coating.

The branched condensation product of terephthalic acid and trimethylhexamethylenediamine is marketed in Germany under the trade name of Trogamid. Because of its irregular structure this nylon is a readily processed transparent amorphous plastic. High melting aromatic polyamides have been produced by the condensation of para-phenylenediamine and terephthaloyl dichloride. More soluble nylons are produced when meta-phenylenediamine is condensed with terephthaloyl dichloride. These polyamides, marketed under the trade name of Nomex, retain their physical properties at temperatures above 300°C.

Nylon-4 has a high moisture absorption similar to that of cotton. It is being produced commercially from pyrrolidone. Nylon-7 is produced in Russia by the polymerization of ω-aminoenanthic acid. The latter is produced from the telomer obtained from ethylene and carbon tetrachloride.

Nylon-11 is being produced from aminoundecanoic acid in Europe. Nylon-12 is also produced in Europe by the ionic polymerization of lauryllactam. Nylons-3 and 9 have been produced on a semicommercial scale.

Polyamides are produced by the condensation of dimerized vegetable-oil acids and polyalkylenepolyamines. These products, marketed under the trade name of Versamid, are used to cure epoxy resins.

12.5 POLYIMIDES

In attempts to produce heat-resistant plastics, polymer chemists have synthesized ladder and semiladder polymers with the hope that random chain cleavage would not necessarily rupture both strands forming the double backbone. Polyimides, which have been known for many years, meet the criterion for semiladder polymers. They are stable in inert atmospheres at 500°C.

As shown in Fig. 12-3, polyimides are produced commercially by a two-stage condensation process. The initial condensation of pyromellitic

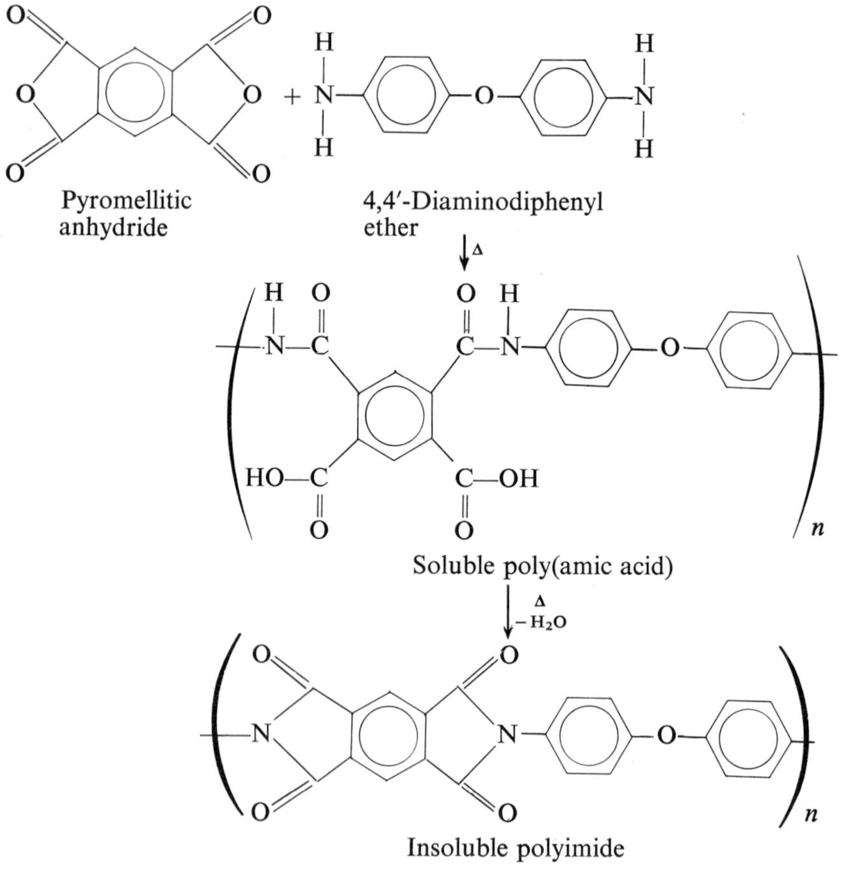

Pyromellitic anhydride

4,4'-Diaminodiphenyl ether

Soluble poly(amic acid)

Insoluble polyimide

Fig. 12-3 Synthesis of a polyimide.

anhydride with 4,4'-diaminodiphenyl ether in anhydrous N,N-dimethylacetamide (DMA) yields a soluble poly(amic acid). The latter is then converted by heating to an insoluble, infusible poly(aromatic imide).

These commercial polyimides were first marketed as heat resistant films and wire enamel under the trade names of H-film and Pyre ML. Billets of these polyimides, which may be machined, now are available under the trade name of Vespel. In some instances, about 15% of graphite is added to the soluble poly(amic acid) before heating in order to produce a filled polymer.

Polyimides are used for seals and valve seats, turbofan engine backup rings, gears, terminal covers, relay actuators, piston rings, and bushings. The properties of a typical machined part of an unfilled polyimide are as follows: Specific gravity 1.43, tensile strength 10,500 psi, flexural strength 15,000 psi, compressive strength 24,000 psi, impact strength 0.9 ft lbs/in. of notch, Rockwell hardness E50, coefficient of expansion 8.0×10^{-4} in./in./°C, heat deflection temperature 680°F, excellent electrical properties, excellent resistance to nonchlorinated solvents, and good resistance to nonoxidizing acids.

A heat resistant *poly(amide-imide)* is also being marketed under the trade name of Torlon. These plastics are also made by a two step process in which the poly(amic acid) is obtained by the condensation of trimellitic anhydride and an aromatic diamine. The soluble poly(amic acid) is converted to an infusible, insoluble poly(amide-imide) by heating.

Fabricated graphite-filled poly(amide-imide) parts are used as nonlubricated and lubricated bearings. The unfilled plastics are used in wire enamels, adhesives and laminates. Considerable information on other heat resistant plastics is supplied in Chap. 15.

12.6 POLYURETHANES

In 1848, Wurtz synthesized low molecular-weight urethanes by the reaction of a monofunctional isocyanate and a monofunctional or monohydric alcohol. As mentioned in Sec. 12.3, German chemists such as Bayer attempted to circumvent the DuPont nylon patents by making polyamides from ω-aminocarboxylic acids. Bayer also used the Wurtz reaction with difunctional reactants to produce nylon-like polyurethanes in 1937.

The first commercial polyurethane was marketed in Germany under the trade names of Igamid U plastics and Perlon U synthetic fibers and bristles. Figure 12-4 shows that the original polyurethanes were produced from the reaction of 1,4-butylene glycol and 1,6-hexamethylene diisocyanate.

During World War II, German chemists were able to use the polyurethane reaction to produce drying oils for paints, adhesives, coatings, elastomers, and both rigid and flexible foams. These polymeric products were described in several reports prepared by American scientists who visited

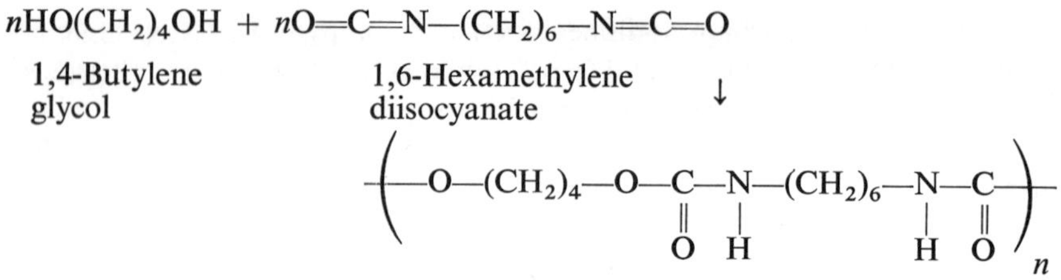

$n\text{HO}(CH_2)_4\text{OH} + n\text{O}=C=N-(CH_2)_6-N=C=O$

1,4-Butylene glycol

1,6-Hexamethylene diisocyanate

↓

$\left(-O-(CH_2)_4-O-\underset{O}{\overset{\|}{C}}-\underset{H}{\overset{|}{N}}-(CH_2)_6-\underset{H}{\overset{|}{N}}-\underset{O}{\overset{\|}{C}}- \right)_n$

Fig. 12-4 Synthesis of a polyurethane.

Germany after the cessation of hostilities. The major interest of American industrialists in the late 1940s was the production of polyurethane foams, previously described in Sec. 4.3. This emphasis on cellular plastics has continued, and almost two billion pounds of polyurethane foam were produced in the USA in 1973.

As illustrated by the structural formulas in Fig. 12-5, the most widely-used diisocyanates today are tolylene diisocyanate (TDI), 4,4-diisocyanatodiphenylmethane (MDI), and polymethylenepolyphenyl isocyanate (PMPI). The most widely-used polyols are polyesters and polyethers with terminal hydroxyl groups. As stated in Sec. 4.3, the original propellant for foams was the carbon dioxide produced by a side reaction between water and the diisocyanate. Fluorochloromethanes (Freon) and other propellants are also used.

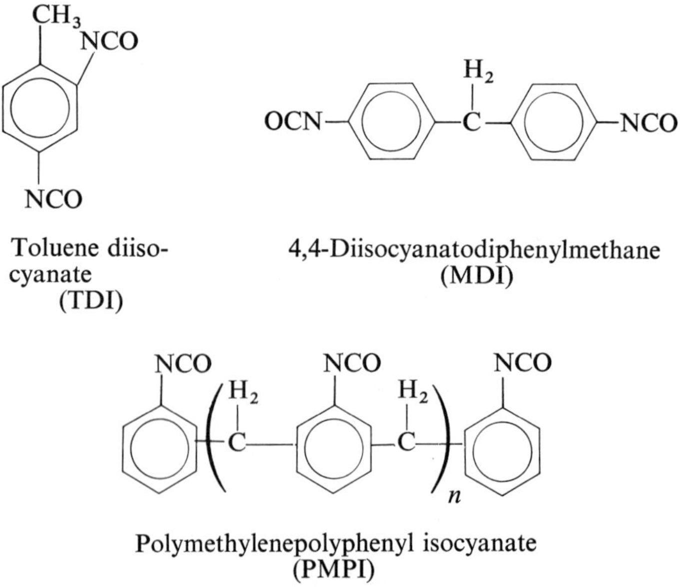

Toluene diisocyanate (TDI)

4,4-Diisocyanatodiphenylmethane (MDI)

Polymethylenepolyphenyl isocyanate (PMPI)

Fig. 12-5 Structural formulas for principal Diisocyanates.

PMPI is widely used as the diisocyanate reactant for the production of polyurethane foams. It is customary to use a polyol with a functionality greater than 2 for the preparation of rigid polyurethane foams. The principal polyols used today are hydroxyl terminated polyethers such as those produced by the reaction of ethylene oxide or propylene oxide with polyhydric compounds such as trimethylolpropane, glycerol, or sorbitol. It is customary to produce a prepolymer by the reaction of the polyol with an excess of the diisocyanate in order to reduce the hazards of handling the latter when producing foams.

The original *flexible polyurethane foams* were produced in Germany under the trade name of Moltopren. These multicellular foams may be made directly from the reaction of a polyol such as the reaction product of propylene oxide and glycerol and a diisocyanate such as TDI in a so-called one shot reaction process. Compounds such as triethylenediamine (Dabco) and dibutyltin dilaurate are often used as catalysts, and silicones are added to control the size of the gas bubbles produced in the foam. Prepolymers made from the same reactants may also be used for the production of flexible polyurethane foams.

Polyurethane coatings are extremely tough and abrasion resistant, but they tend to become yellow. Over 70 million pounds of these coatings were produced in 1974. These coatings may be made by the reaction of diisocyanates with mono or diglycerides of unsaturated acids such as oleic acid.

Coatings may also be produced directly from diisocyanates or prepolymers and polyols or they may be produced by the moisture curing or prepolymers after application. Baked polyurethane coatings may be produced by heating a polyol with an adduct of phenol and a diisocyanate. The latter is called a *blocked isocyanate* since it releases the volatile monofunctional phenol when heated and permits the normal polyurethane reaction to take place with a difunctional glycol.

Adhesive formulations are similar to those used for two pot coating systems. A thermoplastic polyester-urethane elastomer, called Estane, is used for shoe soles. Other elastomers are produced by the reaction of a diisocyanate with flexible polyols having a functionality of greater than 2. Additional diisocyanate reacts during the curing cycle to produce a crosslinked elastomer.

Linear urethane polymers are also widely used as synthetic fibers in Europe. An elastic segmented fiber called Spandex or Lycra consists of blocks of rigid polyurethane and an elastomer. The extruded fibers contain excess isocyanate groups. Those present on the surface are crosslinked by immersion in ethylenediamine.

Polyurethane plastics are linear thermoplastics that may be produced by casting a mixture of prepolymer and polyol or by molding a completely-reacted urethane polymer. Because of the many variables in polyurethane formulations, the properties of these products may be varied from flexible to rigid materials.

12.7 POLYUREAS

Just prior to discovering the reaction of diisocyanates with dihydric alcohols, Bayer prepared polyureas by the reaction of diisocyanates with diamines. The side reaction of water and a diisocyanate yields an unstable carbonic acid that decomposes to form carbon dioxide and an amine. Hence, polyureas are also produced along with polyurethanes in the preparation of foams and by the reaction of amines with freshly extruded spandex fibers. The equation for this reaction was given in Sec. 4.3. The reaction for the formation of polyureas is shown in Fig. 12-6.

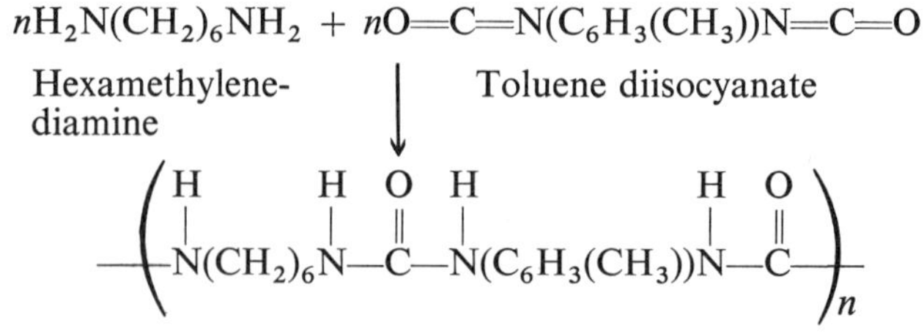

$$n\text{H}_2\text{N}(\text{CH}_2)_6\text{NH}_2 + n\text{O}=\text{C}=\text{N}(\text{C}_6\text{H}_3(\text{CH}_3))\text{N}=\text{C}=\text{O}$$

Hexamethylene-   Toluene diisocyanate
diamine

$$\left(-\underset{|}{\overset{H}{N}}(\text{CH}_2)_6\underset{|}{\overset{H}{N}}-\underset{\|}{\overset{O}{C}}-\text{N}(\text{C}_6\text{H}_3(\text{CH}_3))\underset{|}{\overset{H}{N}}-\underset{\|}{\overset{O}{C}}- \right)_n$$

Polyurea

Fig. 12-6 Synthesis of a polyurea.

Possibly because of the higher cost of amines and the great emphasis on the production of polyurethanes, the polyureas have not been produced commercially in the USA. However, because of their outstanding properties and unusual versatility, they will be considered for commercial production in the future.

BEHAVIORAL OBJECTIVES

While polyimides are related chemically, they vary in physical properties and solubility. The properties of various polyimides and polyurethanes are equally diverse.

After reading this chapter, you should understand the following:

1. Polyacrylamide, which is used as a flocculating agent and scale inhibitor, is produced by the free radical polymerization of acrylamide.

2. Nylon-66, which was the original truly synthetic fiber, is produced by the condensation of adipic acid and hexamethylenediamine. About 10 percent of the annual nylon-66 production is used for plastics.

3. Nylon-6, which is used both as a fiber and as a plastic, is produced by the ionic polymerization of ε-caprolactam. Large articles may be cast by polymerizing this monomer in situ.

4. Nylons with different properties may be produced by the condensation of selected dicarboxylic acids and diamines or by the polymerization of different lactams.

5. Heat resistant polyimides are produced by the condensation of anhydrides of tetracarboxylic acids and diamines. The intermediate soluble poly(amic acids) are converted by heating to insoluble heat resistance plastics.

6. Polyurethanes are produced by the reaction of a diisocyanate and a dihydric alcohol. These versatile polymers may be used as fibers, coatings, cellular products, and moldings.

7. Polyureas are produced by the reaction of a diisocyanate and a diamine.

GLOSSARY

Amide group:
$$-\overset{O}{\underset{\|}{C}}-NH_2$$

DMA: N,N-dimethylacetamide.

Imide group:
$$-\overset{\nearrow O}{C} \diagdown_{NH} \diagup -\overset{\searrow O}{C}$$

Lactam: a cyclic compound consisting of carbon atoms joined to one nitrogen atom. The group adjacent to the nitrogen is a carbonyl (C=O).

PUR: polyurethanes.

Regenerated protein: protein that has been dissolved and reprecipitated.

TDI: tolylene diisocyanate.

Urea group:
$$-\underset{H}{\overset{H}{N}}-\underset{\|}{\overset{}{C}}-\underset{H}{\overset{H}{N}}-$$
$$O$$

Urethane group:
$$-O-\underset{\|}{\overset{}{C}}-\underset{H}{\overset{H}{N}}-$$
$$O$$

QUESTIONS

1. How many carbon atoms are present in the repeating units in nylon-66?
2. Which will be more resistant to moisture, (a) nylon-4 or (b) nylon-7?
3. Why is it difficult to mold polyimides?
4. How would you produce a casting of a polyurethane?
5. How would you produce a polyurethane foam?
6. What reactant is used in place of a glycol for the production of a polyurea?

ANSWERS

1. 6 and 6.
2. (b).
3. Polyimides are not readily softened by heat. Therefore, the soluble poly(amic acid) is used and converted to the polyimide by heating.
4. By reacting a glycol and a diisocyanate in situ.
5. By reacting a glycol and a diisocyanate in the presence of a trace of water.
6. A diamine.

chapter 13

Polynitriles, Polyacetals, and Polyalcohols

13.1 POLYACRYLONITRILE

Acrylonitrile $\left(\mathrm{H_2C{=}\overset{\overset{H}{|}}{C}CN}\right)$ was synthesized by Moureau in 1893, but there was little interest in this monomer prior to the 1930s when German chemists produced an oil-resistant rubber called Buna-N by the copolymerization of butadiene (70) and acrylonitrile (30). The acrylonitrile monomer has been produced commercially from acetylene and from ethylene oxide, but the catalytic ammonoxidation of propylene is now used almost exclusively for the production of more than 1.5 billion pounds of acrylonitrile annually in the USA. The preferred synthesis is shown in Fig. 13-1.

As illustrated in Fig. 13-2, acrylonitrile may be readily polymerized by free-radical polymerization using bulk, solution, suspension, or emulsion techniques. Because of the large difference between the solubility parameter

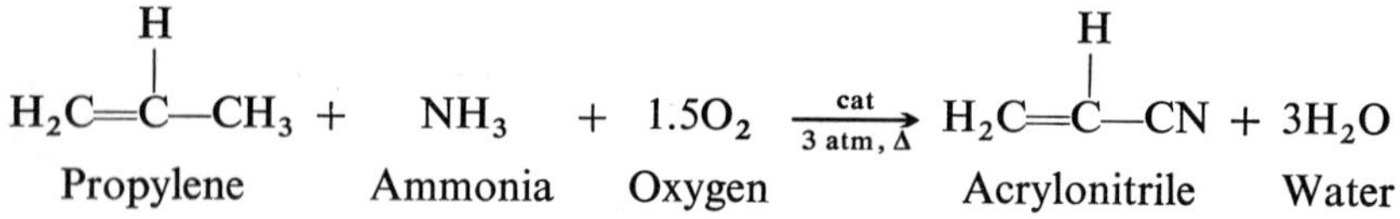

$$H_2C\!=\!\overset{H}{\underset{|}{C}}\!-\!CH_3 + NH_3 + 1.5 O_2 \xrightarrow[3\text{ atm, }\Delta]{\text{cat}} H_2C\!=\!\overset{H}{\underset{|}{C}}\!-\!CN + 3H_2O$$

Propylene   Ammonia   Oxygen   Acrylonitrile   Water

Fig. 13-1   The synthesis of acrylonitrile from propylene.

of the monomer ($\delta = 10.5$) and the polymer ($\delta = 15.4$), the polymer precipitates as a macroradical or a so-called trapped free-radical in its own monomer or in most solvents. The cyano groups form hydrogen bonds with the hydrogen atoms in adjacent chains and thus unplasticized polyacrylonitrile (PAN) like poly(vinyl chloride) (PVC) is difficult to mold or extrude.

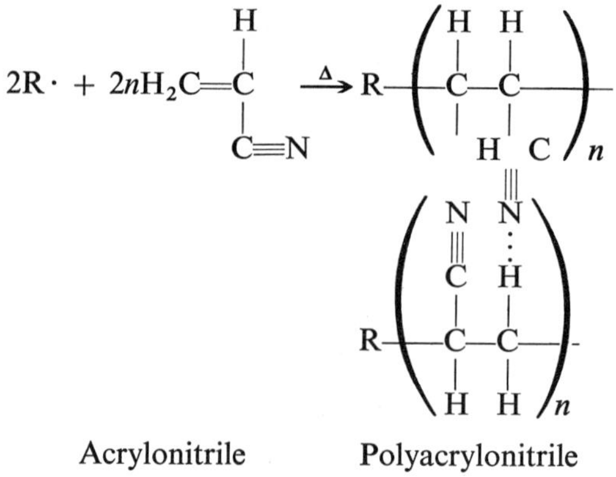

$$2R\cdot + 2nH_2C\!=\!\overset{H}{\underset{C\equiv N}{C}} \xrightarrow{\Delta} R\!-\!\left(\!\!\begin{array}{c}H\;H\\|\;\;|\\C\!-\!C\\|\;\;|\\H\;C\end{array}\!\!\right)_n$$

$$R\!-\!\left(\!\!\begin{array}{c}N\;\;N\\|||\;\;\vdots\\C\;\;H\\|\;\;|\\C\!-\!C\\|\;\;|\\H\;H\end{array}\!\!\right)_n$$

Acrylonitrile   Polyacrylonitrile

Fig. 13-2   Polymerization of acrylonitrile (segments of two chains are shown to demonstrate hydrogen bonding).

It was discovered in the late 1930s that polar solvents with high-solubility parameter values, such as dimethylformamide (DMF), would dissolve PAN. This DMF solution was then passed through spinnerets into a hot chamber or into warm water to produce PAN filaments. These two processes are called *dry* and *wet spinning* respectively.

After plans were made to form a corporation for the production of these fibers, it was found that this polymer, like polypropylene, could not be dyed by standard techniques employed by the textile industry because of the absence of reactive groups. Therefore, reactive dyeing sites were introduced by copolymerizing with a small amount of methyl methacrylate or vinylpyridine $\left(H_2C\!=\!\overset{H}{\underset{|}{C}}\!-\!\underset{N}{\bigcirc}\right)$.

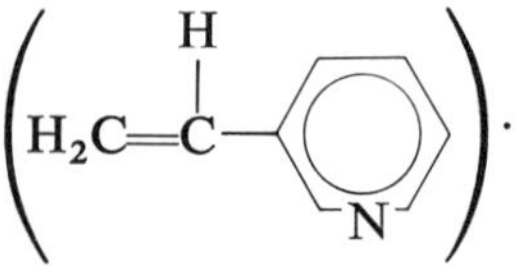

As stated in Chap. 10, fibrous copolymers may also be made by the dry-spinning solutions of copolymers of acrylonitrile and vinyl chloride or vinylidene chloride. These Orlon or Dynel fibers are usually dry spun and are characterized by a *dog bone* cross-section, typical of dry-spun fibers. Since these fibers contain less than 85% acrylonitrile, they are classified as *modacrylic fibers*.

It is customary to use the term *acrylic* to describe all acrylonitrile fibers containing more than 85% acrylonitrile. These are marketed under the trade names of Acrilan and Creslan. The latter is probably a copolymer of acrylonitrile and acrylamide. Another modified acrylic fiber which is said to be an alloy is called Zefran. Bicomponent acrylic fibers resemble wool and are called Orlon Sayelle.

Poly(vinyl chloride) is a crystalline polymer that loses hydrogen chloride when heated. The crystalline acrylonitrile polymer loses hydrogen cyanide in a similar manner when heated. However, instead of forming a dark linear polyene, polyacrylonitrile forms a cyclic polymer that is readily oxidized to a strong, black, heat-resistant ladder polymer, as shown in Fig. 13-3. It is of interest to note that this laddering does not take place when random copolymers of acrylonitrile are thermally decomposed. These ladder polymeric fibers are used as reinforcing agents for epoxy and polyester resins.

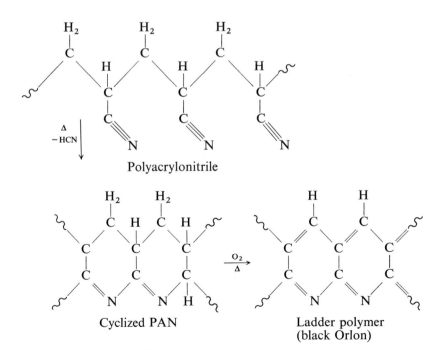

Fig. 13-3 Formation of a ladder polymer by heating PAN.

13.2 COPOLYMERS OF ACRYLONITRILE

Acrylonitrile copolymerizes with many other monomers. As stated in Sec. 13.1, the so-called acrylic and modacrylic fibers are copolymers. Other previously mentioned commercial copolymers of acryonitrile are Buna-N elastomers, ABS copolymers, and styrene-acrylonitrile copolymers.

At least three different companies are producers of terpolymers of acrylonitrile, or methacrylonitrile $\left(H_2C{=}\underset{\underset{CH_3}{|}}{C}{-}CN\right)$, which are being blow molded to form soft drink bottles. These so-called barrier resins may also be molded or extruded. Sheets of these terpolymers may be thermoformed and thermally, solvent, or ultrasonically welded.

The properties of a typical part molded from these acrylonitrile terpolymers are as follows: Specific gravity 1.15, tensile strength 9,000 psi, flexural strength 14,000 psi, compressive strength 12,000 psi, impact strength 2.0 ft lbs/in. of notch, Shore D hardness 80, coefficient of thermal expansion 6.6×10^{-5} in./in./°C, heat-deflection temperature at 66 psi, 165°F, good electrical properties, good resistance to mineral and vegetable oils, ethanol and chlorinated hydrocarbons, and fair resistance to nonoxidizing acids. Blow-molded bottles of this plastic are degraded by sunlight and hence discarded bottles will disintegrate over a period of time. These bottles may also be recycled.

13.3 POLY(PHOSPHONITRILIC CHLORIDE)

Phosphonitrillic chloride (PNCl$_2$) was produced by Stokes in the 19th century and polymerized by Schenck and Rimer in 1924. As shown in Fig. 13-4, this linear high molecular-weight polymer is obtained by heating an intermediate cyclic trimer with phosphorous pentachloride at 350°C.

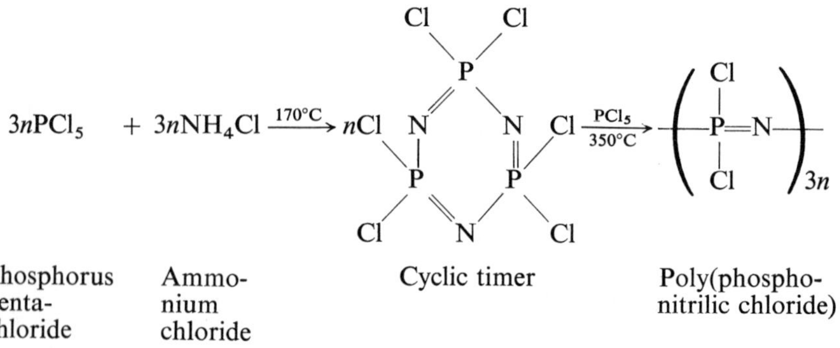

$$3n\text{PCl}_5 + 3n\text{NH}_4\text{Cl} \xrightarrow{170°C} n\text{Cl} \text{ [cyclic trimer] } \text{Cl} \xrightarrow[350°C]{PCl_5} \left(\!\!-\!\!\underset{\underset{Cl}{|}}{\overset{\overset{Cl}{|}}{P}}\!\!=\!\!N\!-\!\right)_{3n}$$

| Phosphorus pentachloride | Ammonium chloride | Cyclic timer | Poly(phosphonitrilic chloride) |

Fig. 13-4 The synthesis of poly(phosphonitrilic chloride).

The original polymers were readily hydrolyzed by water. This deficiency has been overcome by several proprietary techniques, such as heating with alcohols or amines in order to replace some of the pendant chlorine groups. The commercial prepolymer is a fusible, ethanol-soluble powder that, like novolac phenolic resins, is cured by the addition of hexamethylenetetramine and in the presence of magnesia (MgO) at 325–450°F. The dark molded product retains its properties at elevated temperatures. This heat-resistant polymer has been used as a compounding additive for rubber and for molding diamond grinding wheels.

13.4 POLYACETALS

Polymers of formaldehyde were isolated by Butlerov in 1859. The gaseous monomer (formaldehyde) has been used conveniently for more than a century as a 37% aqueous solution, called *formalin*. However, users of formalin, such as biologists, embalmers, and chemists have been plagued by a decrease in the concentration of formalin on standing, resulting from the formation of an insoluble white polymer. Accordingly, it was customary for small-scale users to stabilize formalin by the addition of methanol and for large-scale users to prevent polymer formation by storing formalin at a temperature above 35°C.

The polymer that precipitates in the bottom of a formalin bottle is a highly-branched, low molecular-weight polymer. DuPont chemists were successful in producing a commercial polyformaldehyde in 1959 by the application of modern polymer science. The commercial polymer, called a *polyacetal* or *polyoxymethylene*, is marketed under the trade name of Delrin. It is produced by the cleavage of an extremely-pure, anhydrous, low molecular-weight polymer, such as trioxane. The latter is produced by the alkaline polymerization of formaldehyde.

The pure formaldehyde is cationically polymerized in hexane in the presence of triphenylphosphine $[(C_6H_5)_3P]$ to produce a high molecular-weight polymer with hydroxyl end-groups. A trace of water is added to control the molecular weight.

The thermal or acid degradation of these unstable polymers, called *unzippering*, is prevented by esterifying the hydroxyl or hemiacetal end-groups with acetic anhydride. This so-called *capping process*, which had been used previously by Staudinger to stabilize low molecular-weight polymers of formaldehyde, is also used to stabilize silicone resins. The equation for the production of polyacetal is given in Fig. 13-5.

These highly-crystalline polymers are classified as engineering plastics and are produced at an annual rate of almost one hundred-million pounds. They are available in the form of stabilized molding powders, fibrous glass-filled composites, and as composites of polyacetals and fluoroplastics. Molded and extruded articles of these polymers are used for automotive

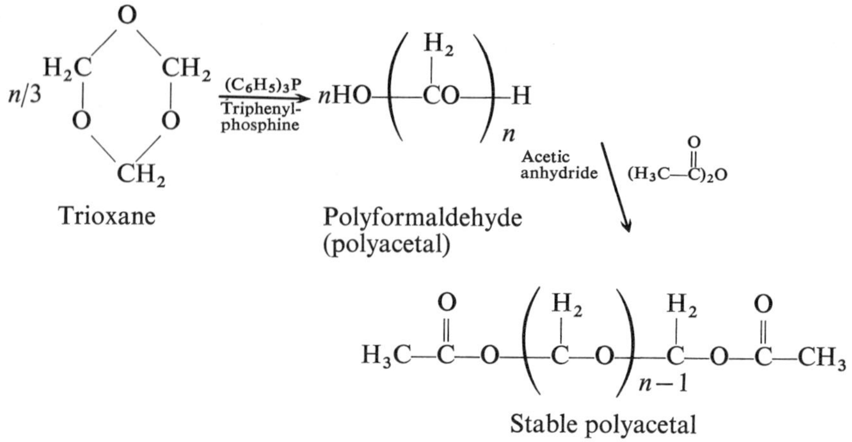

$n/3$ Trioxane $\xrightarrow[\text{Triphenyl-phosphine}]{(C_6H_5)_3P}$ $n\,HO{-}(CO{-}H_2){-}_n{-}H$ Polyformaldehyde (polyacetal)

$\xrightarrow[\text{Acetic anhydride}]{(H_3C-\overset{O}{\underset{\|}{C}})_2O}$

$H_3C-\overset{O}{\underset{\|}{C}}-O-(\overset{H_2}{\underset{|}{C}}-O)_{n-1}-\overset{H_2}{\underset{|}{C}}-O-\overset{O}{\underset{\|}{C}}-CH_3$

Stable polyacetal

Fig. 13-5 Synthesis of polyacetals.

parts and accessories, plumbing fixtures, furniture casters, and components of household appliances.

A typical part molded from this unfilled homopolymer will have the following properties: Specific gravity 1.42, tensile strength 10,000 psi, flexural strength 14,000 psi, compressive strength 18,000 psi, impact resistance 1.4 and 2.3 ft lbs/in. of notch for injection-molded and extruded specimens respectively, Rockwell hardness R120, coefficient of thermal expansion 8.1×10^{-5} in./in./°C, heat deflection temperature at 66 psi 338°F, excellent electrical properties, good resistance to solvents and poor resistance to acids and alkalies.

In addition to use of the capping technique, unstable homopolymers also may be stabilized by reducing the structural orderliness by copolymerization. Thus, the DuPont patents on homopolymers of formaldehyde were circumvented by German and American chemists who produced a commercial copolymer from formaldehyde and small amounts of 1,3-dioxolane 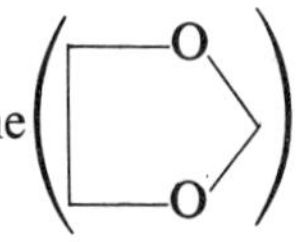 or ethylene oxide. These copolymers, marketed under the trade names of Hostoform and Celcon, must be stabilized before being used as molding powders.

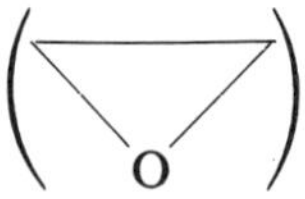

The homopolymer and the copolymer are used for similar applications. Most of the properties of these two plastics are similar, but the heat-deflection temperature of the copolymer at 66 psi is about 20°F less than that of the homopolymer. Homopolymers of acetaldehyde and chloral have also been made in limited quantities.

13.5 POLY(VINYL ALCOHOL)

It is well known by organic chemists that any attempt to produce vinyl alcohol $\left(\begin{array}{c}H\\H_2C\!\!=\!\!COH\end{array}\right)$ by the hydration of acetylene or the hydrolysis of vinyl esters yields the tautomer of this enol viz., acetaldehyde $\left(\begin{array}{c}H\\H_3C\!-\!C\!\!=\!\!O\end{array}\right)$. This rearrangement from the enol to the keto form cannot take place when the polymer of vinyl esters are hydrolyzed. Thus, it was discovered in 1924 that poly(vinyl acetate) (PVAC) could be hydrolyzed by sodium methoxide $(NaOCH)_3$ in methanol to produce poly(vinyl alcohol) (PVAL).

It is customary to stop the hydrolysis when about 85–90% of the acetate groups have been removed. The completely-hydrolyzed polymers of vinyl alcohol are difficult to dissolve because of the high degree of hydrogen bonding between the oxygen atoms and the hydrogen atoms on adjacent chains. This crystalline, atactic, water-soluble polymer is marketed under the trade names of Elvanol and Gelvatol.

Poly(vinyl alcohol) may be plasticized with glycerol or ethylacetamide. The unplasticized polymer is extruded to form fibers that are called Kurlon in Japan. The resistance of these fibers to moisture is increased by immersion in an aqueous solution of formaldehyde. Plasticized, pigmented moldings have been used for oil-resistant gaskets and diaphragms. Extruded plasticized-tubing, marketed under the trade name of Resistoflex, is used for conveying oils and as a flexible conduit for conveying sound.

PVAL films are used for water-soluble packaging such as for soaps or detergents. Since the polymer is a nonionic surfactant and a protective colloid, the disposable package is functional in laundering operations. The surface of the film may be insolubilized by immersion in aqueous formaldehyde, glyoxal, or aluminum alcoholates. Poly(vinyl alcohol) is also used as a textile, a ceramic sizing, a release agent, a thickening agent, and as an ingredient of cosmetics.

13.6 POLY(VINYL ACETALS)

As stated in Sec. 13.5, poly(vinyl alcohol) films and fibers are made water-resistant by immersion in aqueous formaldehyde. As shown in Fig. 13-6, poly(vinyl formal) is produced commercially by the reaction of PVAL and formaldehyde in the presence of acids. This amorphous polymer contains some residual acetate groups because of the incomplete hydrolysis of PVAC and also contains some residual hydroxyl groups because of the random

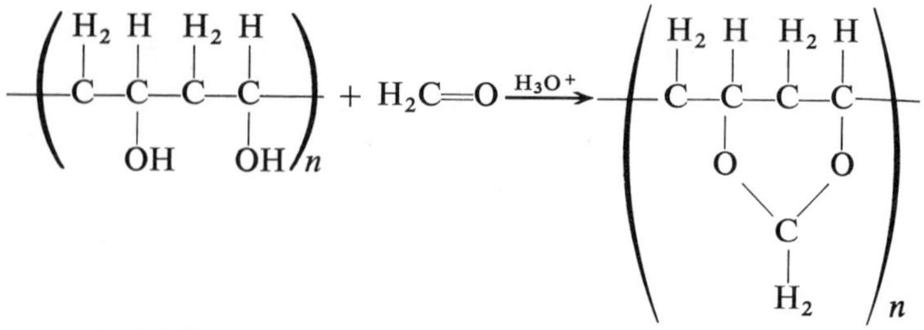

$$\left(\begin{array}{c}H_2\ H\ H_2\ H\\ -C-C-C-C-\\ |\ \ \ \ |\\ OH\ \ OH\end{array}\right)_n + H_2C=O \xrightarrow{H_3O^+} \left(\begin{array}{c}H_2\ H\ H_2\ H\\ -C-C-C-C-\\ |\ \ \ \ |\\ O\ \ \ \ O\\ \diagdown\ \diagup\\ C\\ |\\ H_2\end{array}\right)_n$$

Poly(vinyl alcohol)   Formaldehyde   Poly(vinyl formal)

Fig. 13-6 The synthesis of poly(vinyl formal).

distribution of the hydroxyl groups in PVAL. This acetal reaction takes place between the aldehyde and two hydroxyl groups on alternate carbon atoms. Approximately 13.5% of the hydroxyl groups do not react and are present in *poly(vinyl formal)*.

This tough polymer is marketed under the trade name of Formvar. It is soluble in benzene and chlorinated hydrocarbon solvents. The principal use of this plastic is as a laminating resin and a wire coating. The latter is applied in a cresol solution of a phenolic resin. The baked coating, called Formex, has excellent adhesion to the metal.

The most widely-used poly(vinyl acetal) is produced by a similar reaction between PVAL and butyraldehyde $\left(H(CH_2)_3\overset{H}{C}=O\right)$. It is customary to leave 25% of unreacted hydroxyl groups in the production of this *poly(vinyl butyral)*. This polymer has been marketed since 1936 under the trade names of Butvar, Butacite, and Saflex.

Poly(vinyl butyral) is used in a phosphoric acid solution as a wash primer for metal coatings. This polymer was developed specifically for use as the inner member of laminated safety glass, replacing cellulose nitrate and celluslose acetate, which were not satisfactory. It is customary to plasticize this polymer by the addition of about 35% of a plasticizer, such as dibutyl sebacate. Poly(vinyl butyral) has also been used as a coating for textiles. A poly(vinyl ketal) is produced in Germany by the reaction of PVAL and cyclohexanone.

13.7 POLYMERIC ETHERS

Poly(vinyl ethers) have been produced in Germany since before World War II and in the USA since 1950. As shown in Fig. 13-7, a typical vinyl ether, such as vinyl isobutyl ether, is produced by the Reppe process in which iso-

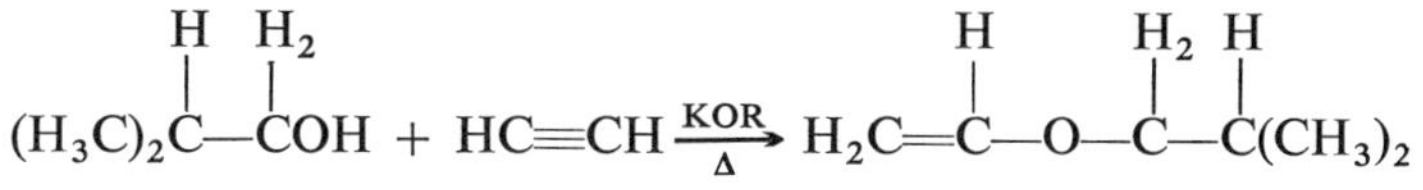

$$(H_3C)_2\overset{H}{\underset{|}{C}}-\overset{H_2}{\underset{|}{C}}OH + HC\equiv CH \xrightarrow[\Delta]{KOR} H_2C=\overset{H}{\underset{|}{C}}-O-\overset{H_2}{\underset{|}{C}}-\overset{H}{\underset{|}{C}}(CH_3)_2$$

Isobutyl alcohol   Acetylene   Vinyl isobutyl ether

Fig. 13-7 Preparation of a typical alkyl ether.

butyl alcohol is added to acetylene in the presence of the potassium salt of the alcohol.

The vinyl ether monomers are polymerized in the presence of Friedel-Crafts catalysts, such as boron trifluoride, as depicted in Fig. 13-8. These polymers are marketed under the trade names of Vinoflex, Resyn, and Oppanol C. They are used as adhesives in pressure-sensitive adhesive tapes and for modifying other polymers, such as rubber. As stated in Chap. 10, the copolymer of vinyl chloride and a vinyl ether is commercially available. Water soluble copolymers with maleic anhydride are marketed under the trade name of Gantrez.

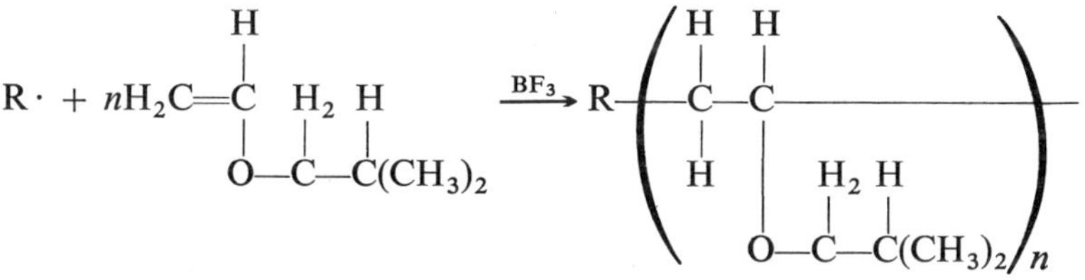

$$R\cdot + nH_2C=\overset{H}{\underset{\underset{O-C-C(CH_3)_2}{|\ \ \ |}}{C}}\ \overset{H_2\ H}{} \xrightarrow{BF_3} R-\left(\overset{H\ \ H}{\underset{\underset{O-C-C(CH_3)_2}{|\ \ \ |}}{\underset{H\ \ H_2\ H}{\underset{|\ \ |}{C-C}}}}\right)_n$$

Vinyl isobutyl ether   Poly(vinyl isobutyl ether)

Fig. 13-8 Polymerization of vinyl Isobutyl ether.

Water-soluble polymers obtained by the ionic polymerization of ethylene oxide were discussed in Sec. 8.6. These polyethers are marketed under the trade names of Carbowax and Polyox and are available in several different molecular-weight ranges. Block copolymers of ethylene oxide and propylene oxide are marketed under the trade name of Pluronics.

13.8 POLYVINYLPYRROLIDONE

N-vinylpyrrolidone, as vinyl alkyl ethers, was also produced in Germany by the Reppe process shortly before World War II. As shown in Fig. 13-9, this monomer is produced by the addition of α-pyrrolidone to acetylene. The former is obtained by the ammonation of the lactone obtained by the dehydrogenation of 1,4-butanediol.

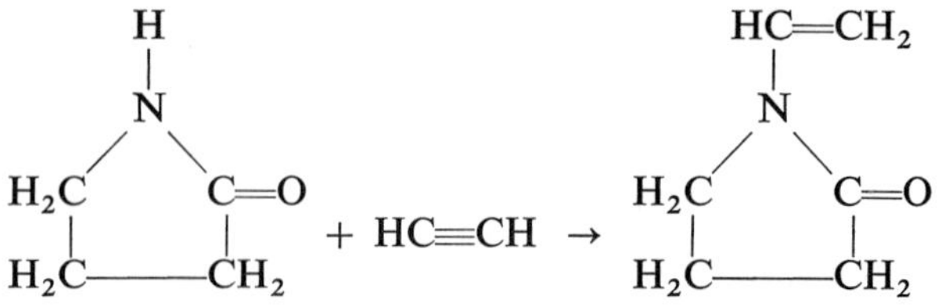

Fig. 13-9 Equation for the synthesis of N-vinylpyrrolidone.

α-pyrrolidone + Acetylene → N-vinylpyrrolidone

Figure 13-10 shows that this water soluble polymer is obtained by the free radical polymerization of an aqueous solution of the monomer.

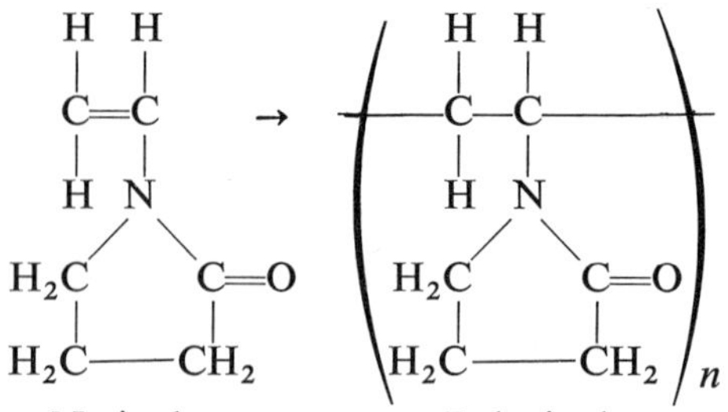

N-vinyl-pyrrolidone → Polyvinyl-pyrrolidone

Fig. 13-10 Equation for the polymerization of N-vinylpyrrolidone (PVP).

During World War II, PVP was used as a blood extender. Whole blood, containing 7% PVP, does not freeze. The polymer also forms a complex with iodine which is used as a germicide. The most widely-used application of PVP is as a hair fixative in aerosol hair sprays. PVP is also used as a dye-stripping agent and as a clarifying agent in the manufacture of beer and wine.

BEHAVIORAL OBJECTIVES

Specialty plastics with properties varying from water soluble to heat resistant water repellant products are produced from the miscellaneous polymers discussed in this chapter. After reading this chapter, you should understand the following:

1. Polyacrylonitrile fibers are obtained by the dry spinning of a solution of a product obtained by the free radical polymerization of acrylonitrile.

Polyacrylonitrile is not used as a plastic but the ladder polymer obtained by the pyrolysis of this polymer is used as a heat resistant polymer.

2. Selected copolymers of acrylonitrile, called barrier resins, are used for the blow molding of bottles for soft drinks.

3. A heat resistant inorganic plastic is produced by the reaction of phosphorus pentachloride and ammonium chloride.

4. Polyacetal is produced by the anionic polymerization of formaldehyde. These engineering plastics are widely used for the molding and extrusion of plastic articles and pipe.

5. Since vinyl alcohol does not exist, PVAL is produced by the hydrolysis of the polymer of the ester of this alcohol, viz. PVAC. PVAL is water soluble.

6. Poly(vinyl acetals) such as poly(vinyl butyral) are produced by the condensation of PVAL and an aldehyde, such as butyraldehyde.

7. Polymers which soften at low temperature and have good adhesive properties are produced by the ionic chain polymerization of vinyl alkyl ethers.

8. A water soluble polymer which has been used as a blood extender is produced by the free radical polymerization of N-vinylpyrrolidone.

GLOSSARY

Barrier resins: usually polymers of acrylonitrile that are used for bottle production.
Capping: the reaction of polymer end groups to prevent depolymerization.
DMF: dimethylformamide.
Dry spinning: the extrusion of a solution of polymers through small holes into a hot chamber.
Hydrogen bonds: forces resulting from the attraction of hydrogen atoms in a molecule with a more electronegative atom such as nitrogen or oxygen in a molecule.
Inorganic: noncarbonaceous compounds.
Ladder polymer: one in which the backbone consists of a double strand much like a ladder structure.
Nitrile group: $\equiv$N
Oxymethylene linkage: —O—CH_2—.
PAN: polyacrylonitrile.
PVAL: Poly(vinyl alcohol).
PVB: Poly(vinyl butyral).
PVF: Poly(vinyl formal).
PVP: Poly(N-vinylpyrrolidone).
Tautomer: an isomer of a ketone.
Wet spinning: the extrusion of a solution of polymer into a nonsolvent.

QUESTIONS

1. Why isn't PAN used as a moldable plastic?
2. What happens when PAN is heated at high temperatures?
3. Why is poly(phosphonitrilic chloride) of great interest?
4. What is the difference between the polymer in the bottom of a formalin bottle and polyacetal?
5. Why is PVAL soluble in water?
6. What is the principal use for poly(vinyl butyral)?
7. Why are poly(vinyl alkyl ethers) not used as molding resins?
8. How could you lower the freezing point of whole blood?

ANSWERS

1. The hydrogen bonds present prevent polymer flow at molding temperatures.
2. It rearranges to produce a heat resistant ladder polymer.
3. Because of its resistance to heat.
4. Molecular weight, molecular weight distribution, heat stability and other important attributes resulting from the application of plastics technology.
5. Because of the presence of many polar hydroxyl groups.
6. PVB is used as the innerliner in automobile windshields.
7. Because of their low softening points.
8. By the addition of PVP.

chapter 14

Starch and Cellulosics

14.1 STARCH

The world's principal food resource is starch. Another polymeric carbohydrate called *cellulose* in the form of its lignin-cellulose composite, which is commonly called *wood*, is the world's most widely-used structural material. Cellulose obtained from cotton is the major fiber used outside the USA. Paper is natural cellulose, and rayon and cellophane are regenerated cellulose. Starch and cellulose account for more than 50% of all organic material.

In addition to its use as food, starch is also used industrially as a sizing and adhesive for paper and textiles. *Dextrin*, produced by the partial degradation of starch, is also used commercially as an adhesive. Starch nitrate, commonly and erroneously called *nitrostarch*, has been used as a coating, but the principal use of this commercial product today is as an explosive.

A detailed discussion of the structure of starch is beyond the scope of this book. However, it is of interest to point out that starch consists of a mixture of two polymeric carbohydrates viz., amylose and amylopectin. The relative amounts of these two types of starch vary with the plant source, but the latter, which is a highly-branched polymer, accounts for about 75% of most industrial starch. Amylose, the minor component, is a linear polymer.

As shown in Fig. 14-1, starch, as most other polymeric carbohydrates, consist of anhydroglucose or glucopyranose units joined together through acetal linkages. These building units may be joined together sterically in two different ways, designated as α and β. Starch is a polymer in which the glucopyranose units are connected by α linkages that are joined to carbon 1 in one unit and carbon 4 in the adjacent unit in the chain. The branches in amylopectin are on carbon 6. This variety has a much higher molecular weight than the unbranched polymer.

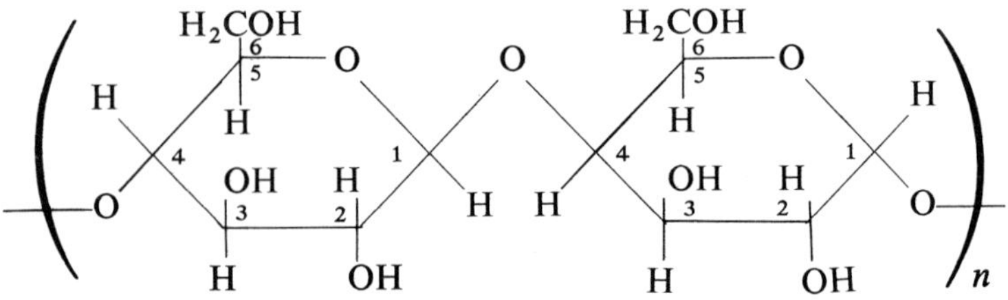

(1,4-α-linkage)

Fig. 14-1 Structure of a repeating segment in the amylose form of starch.

Because of the alpha linkages, the backbone of the starch molecule tends to form helices in which the oxygen atoms and the hydrogen atoms are attracted by intramolecular hydrogen bonds, that is, in the same polymeric chain. These conformations provide space for the iodine molecule in the iodine blue starch test but are not of particular interest to plastics technologists.

14.2 CELLULOSE

In contrast to starch, cellulose is a linear polymer in which the glucopyranose building units or mers are joined by β linkages, as shown in Fig. 14-2. In the preferred conformation for this polymer, the oxygen and hydrogen atoms on adjacent chains are hydrogen bonded. Because of this intermolecular hydrogen bonding, cellulose forms a strong crystalline fiber made up of bundles of polymer chains called *micelles*.

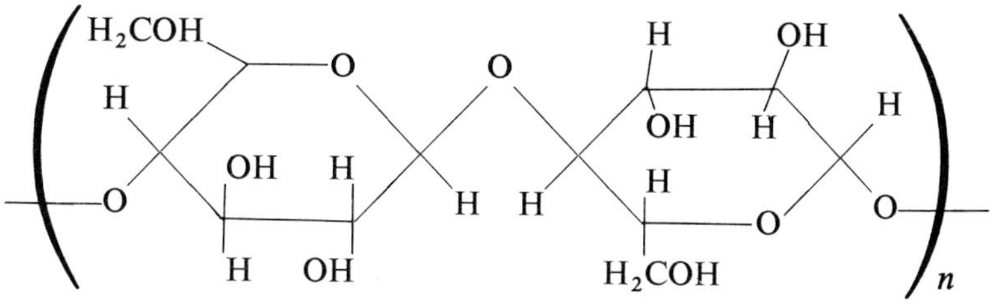

Sec. 14.2 CELLULOSE

Fig. 14-2 Structure of segment of cellulose molecule. (1,4-β-linkage)

Because of its strong intermolecular hydrogen bonds and in spite of the presence of many polar hydroxyl groups, cellulose is insoluble in water. It is a linear polymer, yet it is not a thermoplastic because the total energy required to break these secondary hydrogen bonds is greater than that required to cleave the primary bonds between the carbon atoms in the chain. Thus, the melting point of cellulose is above its decomposition temperature. As a result, cellulose chars and does not flow when subjected to heat and pressure.

The cotton boll is almost pure high molecular-weight cellulose. Wood consists of a mixture of lignin and cellulose. Lignin must be extracted by a reaction with calcium bisulfite, sodium sulfide, or sodium hydroxide. The residual mixture consists of a mixture of cellulose molecules of different molecular weights. This pulp, used primarily for paper making, may be classified on the basis of molecular weight. The highest molecular-weight fraction, called *alpha cellulose*, is insoluble in 17.5% aqueous sodium hydroxide. This fraction and the short linters that remain on the cotton seed are used as a source for regenerated cellulose and cellulose derivatives. Over 100 billion pounds of paper and paperboard were produced in the USA in 1974.

Viscous rayon, which is the major form of regenerated cellulose, has been produced commercially in the USA since 1910. Almost one billion pounds was produced in the USA in 1974, and more than seven billion pounds of this type of rayon was produced annually throughout the world in 1974.

As stated previously, cellulose is insoluble in water because of strong intermolecular hydrogen bonding. However, like other alcohols, it does form a sodium salt when immersed in 17–20% aqueous sodium hydroxide at room temperature. The product, called *alkali cellulose*, reacts readily with carbon disulfide (CS_2) at room temperature to form an alkali-soluble cellulose xanthate. An empirical equation for this reaction is shown in Fig. 14-3.

Rayon is produced by a wet-spinning process in which the viscose solution is forced through spinnerets and the filaments produced are coagulated and converted to cellulose by immersion in an aqueous solution of sulfuric acid and sodium sulfate. Most of the carbon disulfide evolved in this process is recovered and recycled.

$$(C_6H_9O_4(OH))_n + NaOH$$
α-cellulose Sodium hydroxide

$$\downarrow$$

$$(C_6H_9O_4(ONa))_n \xrightarrow{nCS_2} (C_6H_9O_4(O-\overset{\overset{\displaystyle S}{\|}}{C}-S-Na))n$$

Alkali cellulose   Carbon disulfide   Cellulose xanthate

Fig. 14-3 Empirical equation for the production of cellulose xanthate

Cellophane, which is also regenerated cellulose, is produced in a similar manner by forcing the viscose solution through a slit dye. Cellulose sponge is produced by coagulating the viscose solution in an aqueous slurry of sodium sulfate and sulfuric acid. A cellular product is obtained when the undissolved sulfate is extracted by water from the regenerated cellulose product. It is customary to immerse the cellophane sheets in glycerol. About 7% of the glycerol remains in these sheets and serves as a plasticizer. Approximately 500 million pounds of cellophane was produced in the USA in 1974.

The degree of hydrogen bonding in alpha cellulose pulp can also be reduced by immersion in a 75% aqueous zinc chloride solution. Pulp swollen by this technique can be pressed to form sheets. This pressed product, produced under the name of vulcanized fiber, may be postformed to produce various articles, such as helmets and containers.

Rayon has been produced by wet spinning a solution of cellulose in an ammoniacal solution of copper ions. This solution, called *Schweitzer's reagent*, is used in the laboratory for preparing cellulose solutions, but the so-called *Cupra* or *Bemberg* process is no longer competitive with the viscose process for the commercial production of regenerated cellulose.

14.3 INORGANIC ESTERS OF CELLULOSE

Many leaders in the plastics industry cite Hyatt's discovery of Celluloid in 1868 as the beginning of the American plastics industry. It should be pointed out that this plasticized cellulose nitrate is not a synthetic polymer but is a derivative of the natural polymer, cellulose. The name *manmade polymer*, often used to designate such derivatives, is certainly preferable to the much used term artificial. It is important to note that while celluloid and other cellulose derivatives were important thermoplastics. To a large extent they have been replaced by more modern thermoplastics that are produced at an annual rate in excess of one billion pounds. As shown in Table 1-1 in Chap. 1, less than 80 thousand metric tons of cellulosic plastics of all kinds are produced annually in the USA.

Cellulose nitrate, which is erroneously called *nitrocellulose*, was synthesized in the 1830s. However, this inorganic polyester was a laboratory curiosity prior to 1845 when Schönbein accidentally used a mixture of nitric acid and sulfuric acid as a nitrating reagent. Menard, who was serving as Schönbein's assistant, made another serendipitous discovery when he dissolved cellulose nitrate in a mixture of ethanol and ethyl ether in 1846. This solution, called *collodion*, is still available commercially.

Parkesine and celluloid, consisting of filled cellulose nitrate plasticized by camphor, were produced commercially in England and the USA in the late 1860s. The most widely-used cellulose nitrate is a highly-nitrated ester with a degree of substitution (DS) of about 2.8. A look at the cellulose molecule in Fig. 14-2 will show that there are three hydroxyl groups in each glycopyranose building block. All of these are not accessible so that one approaches but does not reach a DS of 3.0 in reactions with cellulose.

The highly-nitrated cellulosic ester, which is used as an explosive, probably accounts in part for the expression that chemistry is exciting! However, it is preferable to leave at least one unreacted hydroxyl group on each anhydroglucose unit when this ester is used as a nonexplosive plastic. Thus, the commercial product used by the plastics and coatings industry is called a dinitrate. It has a solubility parameter of 10.7H and a degree of substitution of about 2.0.

Lower molecular-weight cellulose nitrate, which formerly was widely-used as a lacquer, is obtained by heating the dinitrate in boiling water. Because of its hazardous nature and its poor flow, cellulose nitrate is not injection molded. It is customary to sheet out an ethanol-softened composition on hot rolls and to compress these sheets in a warm press over a long period of time. Thin sheets are obtained by slicing the thick pressed sheets.

An ethanol-softened composition of celluloid may also be extruded to produce rods and tubes. Film is produced by casting a collodion-like solution. Celluloid is a transparent plastic that may be used for the production of brush handles, spectacle frames, knife handles, and ping pong balls.

The original end products were billard balls. Regardless of whether or not celluloid was the first plastic, the techniques now used by the multibillion dollar plastic industry were developed for the fabrication of cellulosic plastics which are now almost defunct. Likewise, the modern synthetic fiber industry was based to a large extent on Chardonnet's extrusion or dry spinning of collodion in the 19th century.

14.4 ORGANIC ESTERS OF CELLULOSE

Cellulose triacetate is less hazardous than celluloid. It was synthesized by Schutzenberger in 1865 by heating cellulose with acetic anhydride. This synthesis was improved by Franchimont who added sulfuric acid to the reactants. Cross and Bevan patented the process for the production of this polymeric ester in 1894.

Unlike the cellulose esters of nitric acid, the degree of substitution could not be controlled by the conditions of the reaction since a product with a DS of about 2.8 was always obtained by the esterification reaction. Since this product was soluble only in expensive chlorinated solvents such as chloroform, its commercial development was retarded. Fortunately, Miles partly saponified this triacetate to produce the acetone-soluble secondary cellulose acetate in 1905.

The cellulose triacetate was actually the first extrudable and injection moldable manmade plastic. The properties of this plastic, which is called Tricel or Arnel, may be modified by plasticization. This plastic continues to be used to a small extent as a molding powder and for the production of tough transparent films. The more soluble and less moisture-resistant cellulose diacetate is also used as a film, as lacquer, and for injection molding. As a result of the energy shortage, the use of cellulosics will increase considerably.

Acetate rayon, which is produced by the dry-spinning of a solution of cellulose diacetate, is widely used as a staple fiber in cigarette filters and as a more expensive rayon with a better "hand" than the xanthate rayon. Over 500 million pounds of acetate rayon were produced in the USA in 1974.

Cellulose acetate butyrate, which is a mixed ester, may be more readily molded than cellulose acetate. It is marketed under the trade name of Tenite. Cellulose acetate butyrate (CAB) extruded pipe was one of the first commercially available plastic pipes and was widely used in gas lines. Cellulose propionate which is marketed under the trade name of Forticel is actually a mixed ester prepared from acetic and propionic acids.

14.5 CELLULOSE ETHERS

Cellulose ethers are produced by the classical Williamson synthesis in which alkali cellulose is heated with an alkyl chloride or sulfate. Thus, in view of the need to replace cellulose nitrate with a less hazardous plastic, it is not surprising that patent applications for the production of ethyl cellulose were filed independently by three plastics chemists in 1912. The degree of substitution may be varied by controlling the conditions of the reaction of alkali cellulose and ethyl chloride (H_5C_2Cl).

Ethylcellulose, marketed under the trade name of Ethocel, is used to a limited extent as a molding powder. This polymeric ether has a DS of 2.5. It is more widely used as a plasticized, strippable, hot-melt, dip coating. Benzyl cellulose was produced in Europe in the 1930s, but its production has been discontinued.

Methylcellulose, with a DS of about 1.7, is produced by the reaction of alkali cellulose and dimethyl sulfate ($(CH_3)_2SO_4$). These polymeric ethers are marketed in several different grades under the trade name of Methocel. Since a moderate amount of substitution reduces the extent of hydrogen bonding, these ethers are soluble in cold water. The products become insoluble when heated and this process is reversible. The temperature of insolubilization, which is called the *thermal gel point*, can by modified by changes in molecular weight, concentration, and the presence of additives. The introduction of a small amount of a large pendant group, such as a hydroxypropyl group, will produce a product that is soluble in hot water.

Hydroxyethylcellulose (Cellosize) and *hydroxypropylcellulose* are also commercially available. These ethers are produced by the reaction of ethylene oxide or propylene oxide with alkali cellulose and are used as thickeners, sizing agents, adhesives, and protective colloids.

Another water-soluble cellulosic ether is produced by the reaction of sodium monochloroacetate and alkali cellulose. These polyethers, which are available in many different grades, are called carboxy methyl cellulose (CMC). The major commercial product with a DS of 0.7 is insoluble in water but soluble in aqueous sodium hydroxide solution.

CMC ethers with a DS of greater than 1.3 are soluble in water and those with a DS of less than 0.3 are insoluble in aqueous sodium hydroxide solutions. Salts of divalent ions such as calcium are insoluble. CMC is used as an additive for detergents, to prevent redeposition of dirt, as a sizing and coating for textiles and paper, as an adhesive, and as a thickener.

Cotton cloth and paper have been made resistant to mildew and bacteria by the reaction of acrylonitrile and alkali cellulose. Cyanoethylcellulose has been produced on a moderate scale and large scale production of this ether is anticipated in the near future.

BEHAVIORAL OBJECTIVES

Because of strong hydrogen bonding, neither starch nor cellulose can be used as processable plastics. However, these forces are lessened in cellulose derivatives which are moldable thermoplastics. After reading this chapter, you should understand the following:

1. The empirical formula of starch is identical to that of cellulose, but neither starch nor its derivatives are used as plastics.

2. The building units in starch and cellulose are identical, but they are arranged differently in space as a result of different acetal linkages. Thus, the starch molecules are helices and those of cellulose are micellular bundles.

3. Since the temperature required to break the intermolecular hydrogen bonds in cellulose exceeds the decomposition temperature, cellulose is not a moldable thermoplastic.

4. Cellulose may be processed when the hydrogen bonds are broken by dissolving in water containing zinc chloride or copper (II) ammonia hydroxide. The hydrogen bonds are also destroyed when the hydroxyls are replaced by other groups such as esters.

5. The polarity of cellulose derivatives increases as the degree of substitution decreases. Thus, cellulose dinitrate is soluble in solvents that are more polar than those that dissolve cellulose trinitrate.

6. Because of its low polarity, cellulose triacetate is not soluble in moderately polar solvents, such as acetone, which will dissolve cellulose dinitrate.

7. Cellulose ethers are produced by the condensation of soda cellulose and an organic chloride. The degree of water solubility of these ethers is related to the number of unreacted hydroxyl groups present in these derivatives.

GLOSSARY

Acetal: an oxygen linkage in starch and cellulose chains.
Amylopectin: highly branched starch molecules.
Amylose: linear starch molecule.
CA: cellulose acetate.
CAB: cellulose acetate butyrate.
Cellophane: regenerated cellulosic sheet.
Celluloid: plasticized cellulose nitrate.
CMC: carboxymethylcellulose.
CN: cellulose nitrate.
Collodion: a solution of cellulose nitrate in ethanol and ethyl ether.
DS: degree of substitution, i.e., the number of derivatives per glucose unit in cellulose.
Ester: the reaction product of a carboxylic acid and an alcohol, i.e., RCOOR.
Ether: a compound in which two carbon atoms are joined by an oxygen atom, i.e., ROR.
Glucopyranose: a six-membered ring present in glucose.
Hyatt, John W.: the inventor of celluloid in 1868.
Iodine test: a purple color observed when iodine is added to starch but not to cellulose.
Manmade: a term which includes truly synthetic polymers as well as derivatives of naturally occurring polymers.
Micelle: a structural unit consisting of bundles of extended polymeric chains.
Nitrocellulose: a trivial name for cellulose nitrate.
Rayon: regenerated cellulosic fibers.
Schweitzer's reagent: copper (II) ammonia hydroxide $(Cu(NH_3)_4(OH)_2)$.

Soda cellulose: a reaction product of cellulose and sodium hydroxide.
Viscose rayon: regenerated cellulose prepared by the acidification of cellulose xanthate.
Xanthate: a salt of a reaction product of carbon disulfide and an alcohol, i.e.,

$$-O-\overset{\overset{\displaystyle S}{\|}}{C}-S^-.$$

QUESTIONS

1. Is starch a monodisperse polymer?
2. What is the difference between the building units of starch and cellulose?
3. Why is it not possible to mold cellulose?
4. Why does glycerol function as a plasticizer for cellophane?
5. Which would be more readily softened by glycerol, (a) cellulose trinitrate or (b) cellulose dinitrate?
6. What are the most widely used moldable cellulosic plastics?
7. Which methyl cellulose is more soluble in water (a) one with a DS of 2.8 or (b) one with a DS of 1.8?
8. Why is a CMC with a DS value of 2.8 more soluble in water than CMC with a DS value of 1.8?

ANSWERS

1. No. It not only occurs in a range of molecular weights but also in two different species, amylose and amylopectin.
2. The building unit of starch is maltose in which two glucose molecules are joined through alpha acetal linkages. The building unit in cellulose is cellobiose in which two glucose molecules are joined through beta acetal linkages.
3. Because of the high temperature required to break the hydrogen bonds.
4. Glycerol is a high boiling liquid having a solubility parameter similar to that of cellulose.
5. (b) because it is more polar.
6. The esters of organic acids such as cellulose di- and triacetate. The hazardous nature of cellulose nitrate restricts its use.
7. (b) because it is more polar.
8. The carboxyl groups in CMC contribute to solubility in water, and the more highly substituted derivative would have more carboxyl groups.

chapter **15**

Ablative and Heat Resistant Plastics

15.1 HEAT RESISTANT PLASTICS

As discussed in Chap. 14, since cellulose nitrate was obviously not suitable for use at elevated temperatures, cellulose acetate was introduced as a replacement for this explosive polymer. Nevertheless, this polymer, like most thermoplastics prior to World War II, was not satisfactory for use at temperatures above 200°F. Fortunately, thermosetting plastics, such as ebonite, phenolic resins, and the so-called amino resins, performed satisfactorily at temperatures as high as 250°F, but they could not be molded by injection-molding techniques that were available at that time.

The challenge to produce an injection-moldable thermoplastic that could withstand the temperature of boiling water was accepted and polymer scientists succeeded in the synthesis of many plastics that met the new specification. A heat resistant PVC was produced by chlorination and isotactic polypropylene, and moldable nylon was also synthesized. Techniques for polymerizing fluoroethenes were discovered and the art of fibrous glass-reinforcement was extended to injection-moldable crystalline thermoplastics.

It is known that stiffening groups in the polymer backbone increase the glass-transition temperature. The ability of polymers to maintain their integrity at elevated temperatures is also associated with geometry permitting good packing and hydrogen bonding that increases the intermolecular forces. Accordingly, several engineering plastics were developed that are useful at temperatures above 100°F. The properties of plastics with moderate heat resistance, such as the polyacetals, chlorinated polyethers, aromatic polyamides, polyimides, poly(aryl ethers), poly(aryl sulfones), polycarbonates, aromatic polyesters, and poly(phenylene sulfides), were outlined in previous chapters.

In spite of the ability of these second or third generation plastics to withstand temperatures previously beyond the limit of most thermoplastics, polymers with completely new temperature-resistant characteristics were essential for the development of the aerospace industry. Hence, two very different types of products were synthesized, i.e., ablative plastics and high temperature resistant plastics.

15.2 ABLATIVE PLASTICS

The word *ablation*, meaning to remove or carry away, was originally coined to describe the process by which the mass of a glacier is reduced by heat. This term, which has been adapted by the space age, is also used to describe the sacrificial pyrolytic degradation or charring of a plastic surface. Thus, the plastic ablator serves as a heat protective device so that while the temperature of the outer surface of a nose cone may be greater than 25,000°F, the inner surface is less than 200°F.

The principle of ablation, permitting nose cone reentry temperatures up to 25,000°F and solid propellant missile propulsion systems to operate at temperatures up to 7,000°F, has been successfully extended to other military and industrial applications. The efficiency of plastic composites as ablators is determined by the Naval Ordinance Laboratories (N.O.L.) alpha rod test in which the flat end of a cylindrical specimen is exposed to the hot gases from an oxyacetylene burner, and both the char rate and temperature are monitored continuously.

Heat shields on large missiles are usually constructed by winding continuous resin-impregnated filament or tape on a rotating mandrel. Rocket nozzles may be compression or gunk molded. The plastic reinforcement may be silica, nylon-66, graphite or asbestos fiber, mat, or cloth. Pressed cork and quartz filler have also been incorporated in phenolic or epoxy resin matrices.

15.3 POLYPHENYLENES

Chemists have not been successful in their attempts to synthesize a tractable poly(p-phenylene), $-(C_6H_4)_n-$. However, like polytetrafluoroethylene, this plastic may be fabricated by sintering techniques, but its use is limited. A benzene-soluble, phenylated polyphenylene, which is stable at 500°C, has also been prepared by heating diethynylbenzene with a bistetracyclone.

15.4 POLYBENZIMIDAZOLE (PBI)

Polybenzimidazole is obtained when a tetrafunctional aromatic amine is heated with a difunctional aromatic carboxylic acid. Thus, a soluble polybenzimidazole is produced when the diphenyl ester of phthalic acid is condensed with a tetraaminobiphenyl in dimethylacetamide (DMAC) at 250°C. The final product is obtained by heating the prepolymer at higher temperatures. This plastic is stable at 500°C in an inert atmosphere. In one modification, an aromatic dialdehyde bisulfite addition product is used in place of the diphenyl phthalate. Both of these polymers are semiladder or stepladder polymers since the double strands are joined by single strands.

15.5 POLYQUINOXALINE

A true ladder polymer that is stable in inert atmospheres at a temperature up to 450°C has been obtained by the heating of aromatic tetraamine and an aromatic tetracarbonyl compound in hexamethylphosphoramide (HMP). The solubility of polyquinoxaline may be improved by using phenylated reactants.

15.6 POLYPYRAZINE

Polypyrazine, which is similar to the more expensive polyquinoxaline, is prepared by heating a bis-α-haloacetyl aromatic compound with ammonia in DMAC. Polypyrazine is stable in temperatures up to 450°C in inert

atmospheres. The solubility of polypyrazine may be improved by adding hydrogen peroxide to a warm solution of this polymer or by using alkyl or phenyl substituted reactants.

15.7 POLYBENZOXAZOLE

Polybenzoxazole is prepared by heating a bisaminophenol and a derivative of a dicarboxylic acid. A polybenzoxazole is stable in an inert atmosphere at 400°C.

15.8 POLYBENZOTHIAZOLE

When a dicarboxylic acid is heated with an aromatic bis-(o-mercaptoamine) in polyphosphoric acid (PPA), a heat resistant polybenzothiazole is produced.

15.9 POLYBENZOXADIAZOLE AND POLYBENZOTRIAZOLE

An aromatic polyoxadiazole is produced by the two-step thermal condensation of a phthaloyl chloride and a hydrazine or a bistetrazole. This light-

sensitive polymer is stable in air at 400°C. It does not melt but is soluble in sulfuric acid or PPA. When the original condensation product is heated with a primary amine, a heat resistant polybenzotriazole is produced.

15.10 POLYHYDANTOIN

Polyhydantoin may be produced by heating a polyisocyanate and a polyglycine. Polyhydantoin is stable at above 300°C.

15.11 POLYKETO POLYTRIKETOIMIDAZOLIDINE

A relatively inexpensive heat resistant polymer may be prepared by heating hydrogen cyanide with an excess of an aromatic diisocyanate. The carbamyl cyanide obtained in the first step will react with another molecule of the diisocyanate to produce an iminoparabamic acid which hydrolyzes to polytriketoimidazolidine.

15.12 PYRRONE

Polybisbenzimidazobenzodipyrrole, which is called a *pyrrone*, is obtained by heating pyromellitic anhydride and diaminobenzidine in DMAC.

15.13 POLYANTHROLINE (BBL)

Polybenzimidazobenzophenanthroline, which is called *BBL*, is obtained by heating naphthalene tetracarboxylic acid with tetraaminobenzene in PPA. BBL is soluble in methane sulfonic acid and is stable in an inert atmosphere at 600°C.

15.14 SPIROPOLYMERS

Intractable spiroketalpolymers, which are stable at 350°C, have been prepared in the laboratory. It is of interest to note that, in contrast to a ladder or semiladder polymer, these have a spiro structure.

In spite of their relative high cost, heat resistant polymers perform a function not duplicated by other materials of construction. These, like many other functional plastics, demonstrate the versatility of polymers and the ingenuity of scientists and technologists associated with all phases of the modern plastics technology.

BEHAVIORAL OBJECTIVES

The use of celluloid, which was the first manmade plastic, was restricted because of its inherent flammability. Some of the synthetic thermoplastics and thermosets which were introduced prior to World War II were more resistant to flames, but they were not useful at the high temperatures encountered in aerospace applications. Hence, new ablative and heat resistant plastics were developed for high temperature service.

After reading this chapter you should understand the following:

1. Thermoplastics with backbones containing stiffening groups and polymers that have strong intermolecular forces are more resistant to elevated temperatures than those with more flexible groups in the chain and those with weak intermolecular forces.
2. Many heat resistant polymers are prepared by condensation of tetrafunctional reactants which produce double stranded chains.
3. Most temperature resistant plastics are prepared by partial condensation reactions and the resulting prepolymers are then heated in a mold to complete the condensation.
4. Many heat resistant polymers consist of double chains shaped like a ladder or like a double helix.

GLOSSARY

Ablation: erosion by heat, i.e., sacrificial pyrolytic degradation.
BBL: polybenzimidazobenzophenanthroline.
BDTA: benzophenone tetracarboxylicdianhydride.
DMA: dimethylacetamide.
HMP: hexamethylphosphoramide.
Intermolecular forces: secondary valence forces such as hydrogen bonds between polymer chains.
Intractable: not softened by heat.
Ladder polymer: one in which the backbone is a double strand with a ladder structure.
PBI: polybenzimidazole.
PPA: polyphosphoric acid.
Sintering: the fusion of a powder by heat and pressure.
Spiro polymer: one having a structure like a double helix.

QUESTIONS

1. Why are ablative plastics not used over long periods of time?
2. Why would a ladder polymer retain its integrity at high temperature better than a single stranded linear polymer?
3. Why are heat resistant plastics shaped while they are in the prepolymer stage?
4. The use of many high temperature plastics is restricted because of high costs, availability, and processing difficulties. How would you design a plastic that would perform at moderately high temperatures?

5. How could failures of plastic structures be prevented, and how could more problems be solved by use of plastics?

ANSWERS

1. Because their performance at high temperatures is dependent on continuous surface erosion.
2. Because the chances of breaking both chains or rungs between rungs or crosslinks are slight and hence a break in the chain does not destroy integrity as it does in a single stranded polymer.
3. Because in contrast to the prepolymers which are soluble in selected solvents and thermoplastic, the final products are insoluble and intractable.
4. Select an available thermoplastic having a relatively high heat deflection temperature. Then reinforce this polymer with a temperature resistant reinforcing fiber as a graphite or boron filament.
5. One solution would be by the use of good plastic technology in the design, selection, application, and use of plastics.

Index

Abbé refractometer, 125
Ablation, 253, 258
Abrasion resistance, 124
ABS, 113, 184, 185, 190
Acetal, 250
Acicular shape, 60
Acid anhydride, 161
Acrilan, 233
Acrylic acid, 206
Acryloid, 207
Acrylamide, 219
Acrylonitrile, 231, 232
Acrylonitrile copolymers, 234
Addition polymerization, 34
Additives, 33
Adipic acid, 219
Advancing of resin, 82, 98, 141
Aging, 23
AIBN, 41, 85, 98
Alkali cellulose, 245

Aliphatic, 161
Alkyds, 150, 152, 161
Allyl plastics, 158
Alpha cellulose, 54, 60, 148, 161, 245
Alternating copolymer, 11, 28
Aluminumtriethyl, 167
Amide group, 229
Amine group, 161
Amino resins, 147, 161
Amorphous, 28, 190
Amylopectin, 244, 250
Amylose, 244, 250
Angstrom unit, 29
Anion, 60
Anisotropic, 50, 60
Antimony oxide, 75
Antioxidants, 64, 66, 78
Antistatic agents, 76, 78
Araldite, 155
Arc resistance, 124

INDEX

Archimedean screw, 105
Arnel, 248
Arrhenius equation, 96, 98
Asbestos, 55, 146
A-stage resin, 45, 60, 141, 161
ASTM, 4, 116, 137
Atactic, 7, 29
ATR, 133
Attenuated total reflectance spectroscopy, 133
Attractions, intermolecular, 15
Azdel, 90
Azobisisobutyronitrile, 41, 85

Backbone, 20
Baekeland, Leo, 141, 161
Bakelite, 29, 146
Banbury mixer, 81
Barrier resin, 241
Barium sulfate, 53
Bavick, 209
BBL, 257, 258
BDTA, 258
Benzophenones, 71
Benzoyl peroxide, 39, 40
Binder, 141
Bingham plastic, 95, 99
Biaxial orientation, 190
Bisphenol A, 162
Block copolymer, 12, 29, 183, 190
Blocking, 78
Blow molding, 109
BMC, 57, 89, 99, 152, 162
Bond dissociation energy, 26
Boron filament, 58
Brabender Plasticorder, 96
Branching, 10
B-stage resin, 46
Bulk density, 137
Bulk factor, 137
Bulk molding compounds, 57, 89
Bulky groups, 21
Butacite, 238
Butvar, 238
Butyl acrylate, 208
Butyl rubber, 13, 29, 122, 175
Butyllithium, 36

CA, 60, 250
CAB, 113, 248, 250
Calcium carbonate, 55
Calendering, 111, 112, 113
Cantilever beam, 137
Capping, 235, 241
Caprolactam, 222
Carbon black, 53

Carbon-carbon bonds, 26
Carbon disulfide, 215, 246
Carbon filaments, 55
Carbonic acid, 216
Carboxymethylcellulose, 249
Cardinol, 162
Carothers' equation, 44, 60
Casein, 218
Casting, 82, 99
Cation exchange resins, 187
Cationic polymerization, 34
Cations, 60, 190
CED, 20, 29
Celanex, 215
Celcon, 236
Cellophane, 246, 250
Cellosize, 249
Cellular plastics, 84, 85
Celluloid, 78, 213, 247, 250
Cellulose, 244, 245
Cellulose acetate, 248
Cellulose nitrate, 247
Cellulose xanthate, 245, 246
Chain reaction, 24
Chain reaction polymerization, 60
Chain transfer, 36, 42, 60, 66, 67
Characterization, 127
Charpy test, 123, 137
Chelate, 78
Chlorinated polyether, 202, 203
Chlorinated polyethylene, 174
Chlorinated rubber, 174
Chromophoric group, 78
Cis, 175
Clay, 51
Cleavage, homolytic, 65
CMC, 249, 250
CN, 250
Cocatalyst, 35
Cohesive energy density, 21
Cold bend test, 119
Cold flow, 13
Colligative properties, 127, 137
Collodion, 247, 250
Colorants, 48
Combustion process, 74
Compounds, 81
Compression molding, 102, 103
Compression ratio, 108
Condensation reaction, 157, 216
Conformations, 5, 29
Coordination complex, 38
Copolymer, 11, 29, 175
Corrosives, effects of, 22, 116
Coumarone, 186
Coumarone-indene resins, 185
CR, 39, 158

INDEX

Creep, 93, 99
Creslan, 233
Cresol, 144, 162
Cross-linking, 13
C-stage resin, 46, 60, 141
Crystallinity, 9
Cyclized rubber, 174, 175

d, 175
DAP, 162
Dash pot, 99
DBP, 208
Degradation of polymers, 65
Degree of polymerization, 19
Delta (δ), 21, 29
Delrin, 235
Density, 118
Density, apparent, 119
Diallyl phthalate, 158
Diatomaceous earth, 60
Dibutyl phthalate, 208
Dielectric constant, 124
Dielectric strength, 124
Differential thermal analysis, 135
Difunctional, 162
Dihydric, 216
Dilatometer, 119
Dimethylacetamide, 225
Dimethylformamide, 232
Dioctyl phthalate, 72, 78
Dipole-dipole interaction, 116
DMA, 225, 229, 258
DMF, 232, 241
Dog bone specimen, 126
DOP, 72, 78, 204
DP, 19, 29
Draw, 113
Drier, 162
Dry spinning, 241
DS, 247, 250
DTA, 135, 137
Dumbbell specimen, 116
DWV, 113
Dynel, 197, 233

e, 99
Ebonite, 174, 175
EEA, 175
Einstein equation, 50, 160
Ekonol, 214
Elastic limit, 137
Elastomer, 29, 175
Elastomers, consumption, 3
Electrical tests, 124, 125
Ellis, Carleton, 53

Eluate, 19
Elution, 19
Elvacet, 213
Embossed, 113
Emulsion polymerization, 39
Encapsulating, 83
End to end distance, 5
Engineering plastics, 191
Environmental stress cracking, 117
EP, 60
EPDM, 170
EPM, 170, 175
Epichlorohydrin, 156
Epoxidized novolacs, 155
Epoxy resin, 47, 154, 155
Estane, 227
Ester, 216, 250
Eta (η), 50, 99, 137
ETFE, 201
Ether, 250
Ethylcellulose, 248
EVA, 171
EVM, 175
Extenders, 49, 73
Extrudate, 113
Extrusion, 105

Fibers, consumption of, 3
Fibrous glass, 55
Fibrous reinforcements, 54
Filament winding process, 89, 90
Fillers, 15, 29, 49, 52
Flame retardants, 74, 78
Flammability tests, 119
Flash, 103, 113
Flexural strength, 123, 137
Flight depth, 108, 113
Flory, Paul, 29
Fluidized bed, 113
Fluoroplastics, 199, 200
Flushed color, 48
Foam, 99
Formaldehyde, 142
Free radical, 23, 30
Free radical polymerization, 38
Freon, 226
Friedel-Crafts' catalyst, 168
Fumed silica, 51

G, 99
Gamma (γ), 99
Gate, 105, 113
GC, 137
Gel coat, 88, 99
Gel permeation chromatography, 19, 131, 132

Gelva, 213
Glass beads, 49, 51
Glass fibers, 55
Glass fillers, 53
Glass mats, 57
Glass transition temperature, 12, 29
Gloss, 126
Glucopyranose, 250
Glycerol, 150
Glyptal, 150
Goodyear, Charles, 13, 45, 71, 175
GP, 104, 113
GPC, 19, 29, 131, 132
Graft copolymer, 12, 29, 191
Graphite reinforcement, 55
GRP, 88
Gutta percha, 172, 175

H-film, 225
Halon, 201
Half lives of initiators, 41
Hardness, 123, 124
Haze, test for, 126
HDPE, 10, 29, 166, 167, 175
Heat deflection temperature, 119, 120
Herschel-Bulkley equation, 96, 99
Hevea braziliensis, 172, 175
Hexa, 162
Hexamethylenediamine, 219
Hexamethylenetetramine, 46, 145
Hindered phenols as antioxidants, 68, 78
HMP, 258
Holographic NDT test, 127
Homopolymer, 175, 204
Hookean, 137
Hooke's law, 92, 99
Hot melt compound, 82
Hot runner molding, 111
Hyatt, John W., 71, 78, 250
Hydrodynamic factor, 50
Hydrogen bonds, 17, 241
Hydrophilic, 78
Hydroxyethylcellulose, 249
Hypalon, 174

Igamid, 225
Ignition temperature, 78
Imide group, 229
Impact strength, 123, 137
Indene, 186
Indene resins, 185
Infrared spectroscopy, 133
Inherent viscosity, 131
Inhibitor, 60
Initiation, 78

Initiators, 34
Injection molding, 109, 110
Inorganic, 241
Intermolecular forces, 258
Intractable, 258
Intrinsic viscosity, 131
Iodine test, 250
Ionomers, 18, 171, 175
IR, 133, 137
Isobutylene, 34, 35
Isotactic, 7, 29, 175
Isotropic, 50, 60
Izod test, 123, 137

Kaolin, 51, 60
Kaurit, 147
Kayser, 133
Kel F, 201
Kelly-Tyson equation, 51
Kienle, 143
Koroseal, 195
Kraton, 184
Kynar, 201

l, 176
Lactam, 222, 229
Ladder polymer, 26, 233, 241, 258
Lamella, 9
Laminates, 91, 99, 141
Land, 113
LDPE, 10, 29, 165, 166, 175
Lead, 108, 113
Lewis acid, 175
Lexan, 211
Light scattering techniques, 129
Lignin, 245
Linear polymer, 29
Log, 99
London forces, 15, 29
Long oil alkyd, 262
Low profile resin, 60
Lubricants, 70, 77
Lucite, 207
Luminous reflectance, 126
Luminous transmittance, 126
Lycra, 227

Macromolecules, 4, 27, 29
Macroradical, 24, 65, 78
Man-made polymers, 247, 250
Mark-Houwink equation, 131, 137
Maxwell model, 93, 94, 99
MDI, 226
Melamine, 150

Melamine plastics, 149
Melt index, 96, 97, 99, 176
Melt viscosity, 95
Melting point, 12, 29
Mer, 11
Merlon, 211
Merriefield synthesis, 187
Metallic fillers, 53
Methacrylic acid, 207
Methyl acrylate, 209, 210
Methylcellulose, 248, 249
Methyl methacrylate, 37, 207
Methylene group, 20, 29
MF, 61, 149, 162
Micelles, 39, 244, 250
Microballoons, 52, 87
Micron, 133
Microspheres, 53
Mineral rubber, 191
$\bar{M}_n$, 19, 129
Modacrylic, 233
Modifier, 42
Mohs' scale of hardness, 51, 60
Mold, positive, 103
Molding powders, 152, 162
Molecular structure, 5
Molecular weight determination, 127
Monodisperse, 18, 126
Mooney equation, 50, 61
Morphology, 30
Mu (μ), 133
Multicellular, 99
Multifoams, 87
$\bar{M}_w$, 19, 29, 30, 129

n, 19
Natta, Giulio, 30
Natural rubber, 172
NDT, 126, 137
Network, 141
Newton's law, 93
Nitrile group, 241
Nitrocellulose, 78, 250
Nitrostarch, 243
NMR, 134, 137
Nomex, 223
Nondestructive testing, 126
Novolac, 45, 61, 144, 145, 162
NQR, 137
NR, 61
Nu ($\bar{\nu}$) wave number, 137
Nuclear magnetic resonance, 134
Nuclear quadrupole resonance, 127
Number average molecular weight, 19, 129
Nylon-4, 223
Nylon-6, 222

Nylon-66, 43, 44, 219, 220, 221, 223
Nylon-610, 223
Nylon-612, 223
Nylon-11, 213
Nylon-12, 213
Nylon-1313, 223

Ochres, 49
Oligomers, 41, 61
One step resin, 141
Oppanol, 239
Organic, 30
Organosol, 196, 204
Orlon, 233
Osmotic pressure, 128, 129
Osmometer, 128
Osmometry, 129
Oxidation, 118
Oxymethylene group, 241

P, 78, 131, 254, 258
PA, 61, 220
Paint, 61
PAN, 61, 233, 241
Paraxylenes, 187
Parylenes, 187
Parison, 113
PC, 61
PE, 142
Pendant group, 30, 162
Pentane interference, 6
Penton, 203
PEPT, 61
Peptizers, 44, 61
Perlon, 225
Perspex, 207
PETP, 215, 216
PF, 30, 162
Pi (π), 138
Phenolic plastics, 140, 141
Phenoxy resins, 189
Phenyl salicylate, 71
Phenylene groups, 19
Phillips' process, 167, 176
Photodegradation, 23
Phthalic acid, 157
Pigments, 48
Pitch, 108, 113
Planck's constant, 133
Plastic, definition, 4, 30
Plasticate, 114
Plasticization, 71, 195
Plasticized PVC, 195, 196
Plasticizer efficiency, 72, 78
Plasticizers, 15, 30, 71, 78, 195, 204

INDEX

Plastics sales, 2
Plastigel, 99, 196
Plastisol, 84, 96, 99, 204
Plate out, 78
Plexigum, 207
Plexiglas, 207
Pliovic, 197
PMMA, 30, 48, 208, 217
PMPI, 226
Poisson's ratio, 92
Polyacetals, 235
Polyacrylamide, 219
Polyacrylonitrile, 232, 233
Polyacrylonitrile cyclization, 26, 233
Polyallomer, 171, 176
Poly(amic acid), 225
Poly(amide-imide), 225
Polyanthroline, 257
Poly(aryl ethers), 187, 188
Poly(arylsulfones), 188
Polybenzimidazole, 254
Polybenzothiazole, 255, 256
Polybenzotriazole, 256
Polybenzoxadiazole, 256
Polybenzoxazole, 255
Poly(butylene terephthalate), 215
Polycarbonates, 211
Polydisperse, 18, 126
Polyenes, 24
Polyester resins, 152, 153, 206
Poly(ethylene oxide), 174
Poly(ethylene terephthalate), 215
Polyethylenes, 164
Poly(4-methylpentene-1), 8, 173
Polyhydantoin, 256
Polyimides, 224
Polyisobutylene, 172
Polymer, 4
Polymer Handbook, 131
Poly(methyl methacrylate), 208
Poly(methyl methacrylate) depolymerization, 26
Polymethylene, 164
Polyolefins, 164
Polyoxymethylene, 255
Poly(p-oxybenzoate), 214
Polypeptides, 183
Poly(phenylene sulfide), 188
Polyphenylenes, 254
Poly(phosphonitrilic chloride), 234
Polypropylene, 168
Polypyrazine, 255
Polyquinoxaline, 254
Polystyrene, 178, 179
Polytetrafluoroethylene, 200
Polytriketoimidazolidine, 256
Polyureas, 228

Polyurethane, 86, 226, 227
Poly(vinyl acetals), 237, 238
Poly(vinyl acetate), 213
Poly(vinyl alcohol), 237
Poly(vinyl butyral), 238
Poly(vinyl chloride), 192, 193
Poly(vinyl formal), 238
Poly(vinyl isobutyl ether), 239
Poly(vinyl pyrrolidone), 239, 240
Poly(vinylidene chloride), 198, 199
Potassium titanate filler, 90
Potting, 83, 99
Powder molding, 111
Power factor, 125
PP, 30, 168, 176
PPA, 258
PPH, 79
PPO, 191
Premix, 99
Prepolymer, 152
Press ratings, 103
Propagation, 39, 78
Protein, regenerated, 229
PS, 180, 191
PSI, 174
PTFE, 200, 204
Pultrusion, 90, 99
PUR, 61, 99, 226, 227, 229
PVAC, 61, 213, 217, 237
PVAL, 61, 237, 241
PVB, 241
PVC, 31, 192, 204, 233
PVCAC, 204
PVDC, 204
PVF, 241
PVP, 240, 241
Pyrolysis GC, 135
Pyrrone, 256

R, 138
R·, 23
Radical, 61
Radius of gyration, 11
Random copolymer, 11
Random flight technique, 6
Rayon, 246, 250
Reduced viscosity, 131
Refraction, index of, 125
Reinforced plastics, 54, 154
Resin, 61
Resole, 141, 162
Resorcinol, 162
Rheology, 91, 99
Rhoplex, 209
Rigid PVC, 194
Root mean square distance, 6

INDEX

Roving, 61
Runner, 104, 114
Ryton, 188

S, 99
Saflex, 238
Saponification, 204
Sapphire filaments, 58
Saran, 198, 199, 204
Saturated, 79
SBR, 183, 191
Schweitzer's reagent, 246, 250
Scratch resistance, 123, 124
Secondary plasticizer, 79
Semon, Waldo, 204
Separan, 219
Sheet molding compound, 51, 57, 89
Shore durometer, 124, 204
Short oil alkyd, 162
Side chain crystallization, 21
Silanol groups, 56
Silica fillers, 51
Silicones, 158, 159, 160
Siloxanes, 159, 162
Sintering, 204, 258
Skeletal formulas, 7, 30
SMC, 51, 57, 61, 89, 152, 162
Soda cellulose, 251
Solubility parameter, 21, 30
Solvent effects, 117
Solvent welding, 191
Spandex, 227
Specific gravity, 118
Specific viscosity, 131
Spectroscopy, 133
Spherulites, 9, 30
Spiropolymers, 258
Sprue, 104, 114
Stabilizer, 69
Stampglas, 90
Standard test conditions, 116
Starch, 243, 244
Staudinger, H., 1, 30
Staudinger index, 131
Step reaction polymerization, 42, 61
Stereoblock copolymers, 170, 176
Stereoregular, 7
Strain, 138
Strand, 61
Stress-strain curves, 120, 121
Structopendant prepolymers, 156
Styrene, 179
Styrene copolymers, 181–183
Styrene, polymerization, 39
Sulfone group, 189
Surface activity, 50

SWP, 204
Syndiotactic, 7, 30
Synergistic effect, 66, 75, 78
Syntactic foam, 61, 87

TAC, 158, 162
Tacticity, 7
Tautomer, 241
TCP, 204
TDI, 85, 227, 229
Tedlar, 202
Teflon, 199, 200
Telogen, 61
Telomerization, 42
Tenite, 215
Tensile strength, 122, 138
Termination of polymerization, 39, 40
Tertiary carbon free radical, 79
Tertiary hydrogen atom, 25
TFE, 199, 200
T_g, 12
TGA, 135, 138
Theoretical plates, 132
Thermal analysis, 135
Thermal tests, 119
Thermoforming, 112
Thermograms, 135
Thermoplastic rubber, 183
Thermoplastics, 14, 30
Thermosets, 14, 22, 30
Thiourea, 148
THF, 132
Titanium tetrachloride, 167
Toggle, 114
Tolylyl diisocyanate, 85, 227
Tool, 103
Torlon, 225
TPX, 173, 176
Trans, 176
Transfer mold, 104, 114
Transition zone, 106
Transitions, 12
Transport zone, 106
Triallyl cyanurate, 158
Tricresyl phosphate, 73
Type C glass, 61
Type E glass, 61
Tyrin, 174

UF, 61, 162
UHMWPE, 167, 176
Ultra-high molecular weight PE, 167
Ultramarine blue, 49
Ultraviolet light degradation, 25
Unsaturated, 79, 162

Unsaturated polyesters, 153
Unzippering, 235
Urea group, 229
Urea resins, 147
Urethane group, 229
UV, 79
UV stabilizer, 70, 71

Valox, 215
Vapor phase osmometry, 129
VCM, 71, 204
Vespel, 225
Vicat test, 119
Vinoflex, 197
Vinyl acetate, 213
Vinyl chloride, 193
Vinyl chloride copolymers, 197, 198
Vinyl fluoride, 202
Vinyl plastic, 204
Vinylpyrrolidone, 240
Vinylidene chloride, 198
Vinylidene fluoride, 202
Vinylite, 197, 204
Viscoelastic, 91, 92, 99
Viscometry, 130, 131
Viscose rayon, 251

Viscosity, 99
Viscosity number, 131
Voigt model, 94, 95, 99
Volume, excluded, 6
Volume, hydrodynamic, 132
Volume resistivity, 125
VYHH, 197
VYNW, 197

Wave number, 133
Weathering tests, 117
Weight, average molecular, 19, 129
Wet spinning, 241
Whiskers, 57, 61
Wood flour, 53, 141, 146, 148, 162

Xanthate, 251

Yield point, 138

Ziegler, Karl, 61
Ziegler-Natta catalyst, 34, 61, 167, 176
Zinc oxide, 53
Zipper reaction, 69, 78

SPRING '83